Fishes

Guidelines for the breeding, care, and management of laboratory animals

A Report of the
Subcommittee on Fish Standards
Committee on Standards
Institute of Laboratory Animal Resources
National Research Council

NATIONAL ACADEMY OF SCIENCES Washington, D.C. | 1974

NOTICE: The project that is the subject of this report was approved by the Governing Board of the National Research Council, acting in behalf of the National Academy of Sciences. Such approval reflects the Board's judgment that the project is of national importance and appropriate with respect to both the purposes and resources of the National Research Council.

The members of the committee selected to undertake this project and prepare this report were chosen for recognized scholarly competence and with due consideration for the balance of disciplines appropriate to the project. Responsibility for the detailed aspects of this report rests with that committee.

Each report issuing from a study committee of the National Research Council is reviewed by an independent group of qualified individuals according to procedures established and monitored by the Report Review Committee of the National Academy of Sciences. Distribution of the report is approved, by the President of the Academy, upon satisfactory completion of the review process.

This study was supported in part by Contract PH43-64-44 with the Drug Research and Development, Division of Cancer Treatment, National Cancer Institute, and the Animal Resources Branch, National Institutes of Health, U.S. Public Health Service; Contract AT (11-1)-3369 with the Atomic Energy Commission; Contract N00014-67-A-0244-0016 with the Office of Naval Research, U.S. Army Medical Research and Development Command, and U.S. Air Force; Contract 12-16-140-155-91 with the Animal and Plant Health Inspection Service, U.S. Department of Agriculture; Contract NSF-C310, Task Order 173, with the National Science Foundation; Grant RC-10 from the American Cancer Society, Inc.; and contributions from pharmaceutical companies and other industries.

Library of Congress Cataloging in Publication Data

National Research Council. Institute of Laboratory Animal Resources. Subcommittee on Fish Standards. Fishes: guidelines for the breeding, care, and management of laboratory animals; a report.

Bibliography: p.
1. Fish as laboratory animals. I. Title.
II. Title: Guidelines for the breeding, care, and management of laboratory animals.
SF407.F5N37 1974   639'.34   74-3462
ISBN 0-309-02213-4

Available from
Printing and Publishing Office, National Academy of Sciences
2101 Constitution Avenue, N.W., Washington, D.C. 20418

Printed in the United States of America

Preface

Two thirds of the earth's surface is covered with water inhabited by invertebrate and vertebrate aquatic animals that are important as sources of food for people and domestic animals. Of these, fishes are the most abundant and diversified vertebrates. In research they have certain advantages over other animals since many aspects of their physiology can be manipulated by changing the water temperature. Their adaptation to a wide range of environmental conditions and their susceptibility to chemical and physical agents make them highly useful in water pollution studies.

Terrestrial animals breathe atmospheric oxygen that is uniformly distributed over the earth at about 20 percent of the air. Fishes can live in water containing dissolved oxygen in quantities as low as 1 part per million (ppm) to saturation, about 14 ppm. Some species can tolerate water temperature differences of only a few degrees from their optimum; other fishes can exist in temperatures varying from just above freezing to 35 °C (95 °F). They also may inhabit water containing only traces to a high percentage of dissolved mineral salts.

Because increasing numbers of fishes are used as experimental animals, it is important that certain standards be established to facilitate the maintenance of fishes in research and to enhance the reproducibility of results. This book contains basic information on sources, maintenance, handling, and use of fishes as experimental animals.

SUBCOMMITTEE ON FISH STANDARDS

S. F. SNIESZKO, *Chairman*
HERBERT R. AXELROD
R. E. GOSSINGTON
KLAUS D. KALLMAN
STEWART McCONNELL
CHARLES R. WALKER

Contents

I General Requirements of Experimental Fishes

The methods for assessing the quality of fishes used in research are not well established, and information on their acquisition is lacking. Most fish culture techniques now in use are for mass production of fishes and do not incorporate the special care necessary for quality research fishes. Information on standard procedures is needed to guide suppliers of quality experimental fishes to better ensure reproducible results (Henderson and Tarzwell, 1957; Lennon and Walker, 1964; Lennon, 1967; Brauhn and Schoettger, in press).

It is essential for research workers not familiar with fishes to understand that the quality of the environment is critical to the well-being of the fishes (see Chapter VIII, Section B). Fishes are ectothermic animals; their body temperature approximates that of the water in which they live. Metabolism and growth rate are strongly influenced by temperature. The temperature of water used for holding fishes varies within a wide range depending upon the species; however, above their optimum temperature range, fishes may be under stress; below the optimum range, their metabolism is slow, they have little demand for food, and they do not respond to antigenic stimulation (Snieszko, 1970; Anderson, 1974).

According to species, fishes vary in their tolerance to saline concentrations in the water. While some species can live only in fresh water or seawater, others readily make the adjustment to changes from fresh to seawater or vice versa. Euryhaline fishes have a wide physiological tolerance for dissolved salt, while stenohaline have none. Diadromous fishes readily adjust to the change in osmolarity, as evidenced by their migrations from fresh to salt water.

The procedures used in collecting fishes from nature or rearing ponds often produce injuries and stress that may not be obvious at the time. The results of experiments conducted with such fishes are often questionable.

Laboratories routinely using large numbers of fishes often place standing orders to obtain them from private or government-operated fish farms. However, even the best fish farms do not now attain the high standards of quality expected of mammals raised as laboratory animals. As a result, some users have to resort to breeding and raising their own fishes to achieve the desired quality.

Many trout hatcheries and trout farms produce relatively high quality fishes. Rainbow trout are being maintained and bred successfully on a completely defined diet (Halver, 1972). Some fish farms are producing various species of minnows, goldfish, catfish, bluegills, and tropical fishes on a large scale. Selected marine fishes are cultured in enclosures in sheltered coastal areas in the United States, western Europe, the Philippines, and Japan (Bardach *et al.*, 1972; Richardson, 1972; Heighway, 1973).

Only a small number of fish species have been inbred under controlled conditions. Among those that are maintained as pure genetic lines are strains of platyfishes and swordtails (Kallman, 1970; Kallman and Atz, 1966). Many other aquarium fishes are also cultured in fish farms or can be bred and raised in the laboratory (Axelrod and Shaw, 1967; Agranoff *et al.*, 1971; Axelrod, 1971).

II Selection of Laboratory Fishes

Before making a selection of types or species of fishes suitable for specific research objectives, the investigator should evaluate the physical facilities available to him. Fishes have diverse habitat requirements and different species require varying conditions. To facilitate selection, some basic physiological requirements of fishes are listed below.

- Availability of Dissolved Oxygen. The concentration of dissolved oxygen available to fishes is often critical. The dissolved oxygen in water at temperatures of 10–30 °C (50–86 °F) at sea level varies from 11 to 8 ppm. Certain fishes have high oxygen requirements. Salmonids in particular cannot live in water containing less than 5 ppm of oxygen. In other words, only half of the dissolved oxygen is available for these species. The rate of oxygen depletion is also directly related to the number of fish per unit volume of water.

Many fishes can survive in water containing as little as 3–5 ppm of dissolved oxygen. However, these fishes usually respond to an increase in the oxygen concentration with a higher rate of metabolism and accelerated growth. An increase of dissolved oxygen by 2–3 ppm above the critical level may double the rate of growth.

- Carbon Dioxide. Carbon dioxide, one of the major waste products of fishes, seriously interferes with the oxygenation of hemoglobin. It is present in the form of carbonic acid. An increase in dissolved carbon dioxide raises the critical level of oxygen in the water, below which oxygen becomes unavailable for fishes. Elimination of carbon dioxide and the replenishment of oxygen takes place only at the surface (air–water interface). Oxygen can be greatly increased by conducting a vigorous stream of air through the water.
- Ammonia. Ammonia is the main nitrogenous waste product of

fishes and is secreted by the kidney and the gills. Ammonia is highly soluble in water and is, therefore, not eliminated from the water through aeration. At a pH lower than 7 most of the ammonia is present in the ionized form. At pH 8 or above, most of it is present in the nonionized form. Both forms of ammonia are highly toxic to fishes, but the nonionized form is much more toxic. Many species cannot tolerate more than 1 ppm free ammonia in water. For salmonids, more than 0.2 ppm is undesirable.

Ammonia in a well-maintained aquarium is oxidized by nitrifying bacteria to much less toxic nitrite and then to nitrate, which is less toxic than other forms of nitrogen.

• Temperature. Fishes of various species have different optimum temperatures ranging from about 10 to 35 °C (50 to 95 °F). Some species are less tolerant to temperature changes than others, but even the fishes with wide tolerance must be slowly adapted to changes that are greater than 5 °C (10 °F). At higher temperatures fishes metabolize faster and the quantity of dissolved oxygen is lower; therefore, the rate of aeration of the water should increase proportionally to the temperature.

• Salinity. Different species of fishes exist at various temperatures and salinities of water. Fishes can be divided according to salinity tolerance into freshwater and marine species. Estuarine fishes such as *Fundulus* are euryhaline and tolerate a wide range of salinity. Anadromous and catadromous fishes spend part of their lives in fresh water and part in marine environments. Some anadromous fishes, such as salmons, may become landlocked in a freshwater environment or cultured in a marine environment. When a high degree of reproducibility of experimental results is required, it is customary to maintain fishes in deionized or distilled water to which minerals are added at a standardized concentration (Lennon and Walker, 1964; see Chapter VIII).

Research workers must be aware that usually more factors must be standardized during experiments with fishes than with terrestrial, warm-blooded vertebrates. *Without the maintenance of uniformity of dissolved oxygen, carbon dioxide, ammonia, nitrate, temperature, pH, and mineral content of water, it is not possible to reproduce the results of experiments in which fishes serve as experimental animals.*

• Availability. Laboratory fishes should be readily available throughout the year. They must be capable of being maintained and able to reproduce under laboratory conditions (see Chapters VIII and X). In contrast to fishes collected from native habitats, cultured species offer the advantage of known history, age, nutrition, genetics, and health. The fishes listed below by ecologic type are not the only species recommended, but are examples of those suitable for experimental use (Nigrelli, 1953; Pickford and Atz, 1957; Lennon and Walker, 1964; Agranoff *et al.*, 1971; Binkowski, 1972; Umbreit and Ordal, 1972).

A. FRESHWATER

Other than for breeding purposes, the average laboratory may be able to maintain the following species under satisfactory conditions:

Rainbow trout	*Salmo gairdneri*
Channel catfish	*Ictalurus punctatus*
Bluegill	*Lepomis macrochirus*
Fathead minnow	*Pimephales promelas*
Goldfish	*Carassius auratus*

B. DIADROMOUS

Of the diadromous fishes the most frequently used in nutritional and physiological research are eels and the Atlantic and Pacific salmons.

C. ESTUARINE AND MARINE

Most of the estuarine and marine fishes used for laboratory research are wild fishes obtained at cyclic intervals. The maintenance of these fishes in the small-scale aquarium necessitates development of a working knowledge of the entire ecosystem. The most extensively used estuarine species is the mummichog (*Fundulus heteroclitus*).

The following species of estuarine and marine fishes have been maintained in a closed, recirculated, artificial seawater system for varying lengths of time:

NATIVE	
Striped mullet	*Mugil cephalus*
Red drum	*Sciaenops ocellata*
Atlantic croaker	*Micropogon undulatus*
Southern flounder	*Paralichthys lethostigma*
Pinfish	*Lagodon rhomboides*
Sheepshead minnow	*Cyprinodon variegatus*
Mummichog	*Fundulus heteroclitus*
Gulf killifish	*Fundulus grandis*
Tidewater silverside	*Menidia beryllina*
Oyster toadfish	*Opsanus tau*
Spot	*Leiostomus xanthurus* Lacépède

EXOTIC

Wrasses	Labridae
Parrotfishes	Scaridae

D. TROPICAL

There are many sources of healthy tropical fishes. Foremost among these is the Genetics Laboratory associated with the New York Zoological Society and located at the Osborn Laboratories of Marine Sciences, Brooklyn, New York. These laboratories have many inbred, disease-free strains of swordtails and platyfishes suitable for research use. Also, there are fish farms in Florida raising tropical fishes, some of which are genetically uniform.

Although some can be bred only with difficulty, the following species are listed as examples of satisfactory experimental fishes:

FRESHWATER LIVEBEARERS

Guppy	*Poecilia reticulata*
Southern platyfish	*Xiphophorus maculatus*
Green swordtail	*Xiphophorus helleri*
Montezuma swordtail	*Xiphophorus montezumae*
Mosquitofish	*Gambusia affinis*
Least killifish	*Heterandria formosa*
Amazon molly	*Poecilia formosa*
Sailfin molly	*Poecilia latipinna*
Shortfin molly	*Poecilia mexicana* Steindachner

FRESHWATER EGGLAYERS

Zebra danio	*Brachydanio rerio*
Medaka	*Oryzias latipes*
Siamese fighting fish	*Betta splendens*
Paradisefish	*Macropodus opercularis*
Dwarf gourami	*Colisa lalia*
Blue gourami	*Trichogaster trichopterus*
Blackchin mouthbrooder	*Tilapia melanotheron*
Blue acara	*Aequidens pulcher*
Mexican tetra	*Astyanax mexicanus*
Electric eel	*Electrophorus electricus*

III Guidelines for Uniformity

A. TAXONOMY AND IDENTIFICATION

Experimental animals should be properly identified by genus, species, and subspecies or strain. Some specimens should be preserved for future reference in 70 percent isopropyl alcohol or 10 percent buffered formalin and deposited in a permanent institution or laboratory. The misidentification of even so-called common, well-known species has lessened the value of many excellent experiments (Bailey, 1970).

B. GEOGRAPHICAL ORIGIN

The origin of commercial fishes should be suspect since dealers' shipments may comprise mixed lots obtained from two or more locations. Important differences often exist in physiological, structural, and behavioral parameters between different populations of the same species. These differences may be environmentally or genetically controlled. Changes in certain parameters due to selection over a period of years, or even decades, may take place in the same habitat. The interpretation of experimental results (e.g., different findings of two laboratories, inability to repeat experiment) may be more meaningful when the precise source or strain of the experimental animal is known. Intraspecific differences can then perhaps be correlated with ecology. Experimental results are easier to reproduce when animals from the same habitat are used repeatedly.

Descriptions of locations where native fishes are taken, such as eastern North America or Central America are of little value. A permanent, written record should be kept as to when, where, and by whom the fishes were obtained. Information must be provided as to river systems, state, county, and the distance to the nearest recognizable, permanent geographical landmark. Additionally, fishes can be procured from established, reliable hatcheries that keep an adequate record of the history of the stock.

C. SIZE

The most convenient size of fishes for testing the acute effect of chemicals added to water is up to 50 mm (2 in.) in length. In long-range (chronic) tests fishes of various sizes and various ages are usually used. In physiological experimentation, fishes 0.3-0.5 m (10-20 in.) long are often preferred for confinement in sophisticated apparatus and enclosures. In immunological research studies fishes 255-510 g (9-18 oz) are required to obtain 5-10 ml (0.2-0.3 fl oz) of serum.

The biomass ratio of fishes to water is very important in physiological and toxicological research. Even if the same ratio of fish weight to volume of water is maintained, results of experiments may depend upon the size of individual fish. The body surface of the fish plays an important role during exposure to toxic chemicals.

D. AGE

Since important physiological changes take place during the life span of a fish, individuals of known age are required for many experiments. Age of fishes can occasionally be determined by microscopic examination of the scales; however, this requires special equipment and experience. For small tropical fishes this method is not effective.

Juveniles should be collected and held together to ensure a fairly uniform group with respect to size and age. The most reliable way to obtain fishes of a known age is to breed and raise them under controlled conditions. Fishes from commercial sources may have mixed age-groups in the shipment.

E. SEX

Sex in juvenile fishes is difficult to determine, but in certain groups mature males and females can readily be distinguished from each other by the external secondary sex characteristics (differences in coloration, fin structure, genital papilae, etc.). Mature tropical fishes are usually easily sexed; native cold-water fishes are not (unless determined during spawning activity). In certain species sex can be ascertained only by autopsy or by behavior patterns. References on fish taxonomy should be consulted for correct sex identification.

F. GENETIC CONSIDERATIONS

In the past, little attention has been paid to the genetic background of fishes used in research, although virtually every characteristic is ultimately influenced by the genotype. No two individuals of a natural population are genetically identical, even though all members of the population may share a common gene pool. Testing an adequate sample of a population will yield an average value characteristic for this population. Such tests can be repeated with identical results as long as the population does not change (selection) or become contaminated by individuals from other gene pools. Somewhat different values may be obtained from experiments involving a different population of the same species. Examples of pure populations needed for research are provided by Martin's (1968) studies on intraspecific variations in osmotic abilities of *Cyprinodon variegatus* and Kallman's (1970) pigment pattern analyses in *Xiphophorus maculatus*.

Inbred stocks are required for experiments in which reproducible responses to be measured and compared with other experimental results are needed. Inbred stocks are genetically homozygous and are the product of at least 20 generations of brother-to-sister matings. For most uniform results F_1 hybrids between two inbred strains are often more useful than the parental stocks, because inbreeding may increase the variability in response to a certain treatment (Lerner, 1954; Lindsey, 1969).

To minimize experimental variation and ensure known lineage, hatcheries maintaining stocks several generations removed from wild populations should be considered as the principal or preferred source of laboratory fishes.

G. HEALTH CONDITIONS

Fishes must be free from any overt signs or symptoms of diseases and parasites. Fishes are often affected with mild chronic diseases or are asymptomatic carriers. One of the most important criteria in determining health condition is that the fishes be free from specified diseases and parasites at the production facility for a period of at least one year. Fishes obtained from wild conditions or from an uncertified primary producer or collector should be maintained in quarantine or isolation until determined healthy (see Chapter VII).

H. CONTAMINATION WITH CHEMICALS

Most fishes contain residues of pesticides and other chemicals (Henderson *et al.*, 1959; Brauhn and Schoettger, in press). Laboratory fishes to be used in physiological and toxicological research should be examined for contaminants such as pesticides, heavy metals, and industrial chemicals preferably before shipment (Brauhn and Schoettger, in press). Channel catfish from private hatcheries in the south-central United States were found to have residues of nine pesticides. Each pesticide or industrial contaminant was present at a different level; some pesticides were present in all fishes examined. Compounds such as toxaphene, polychlorinated biphenyls (PCB's), phthalate esters, dieldrin, and endrin were commonly present (Bureau of Sport Fisheries, 1972).

The procedure of securing fishes with low pesticide residues for pesticide research is twofold. First, each group of fishes from the supplying hatchery should be analyzed before shipment. Second, approximately 2 weeks prior to the research, a 20-g (0.7-oz) sample of fishes from the group reaching the desired size at the supplying hatchery should be analyzed for pesticide residues. This preshipment check helps researchers decide whether the residues present will interfere significantly with the planned research. If the planned research will be jeopardized by the use of a particular group of fishes, the shipment should be cancelled and an alternate source of supply sought.

IV Uses of Laboratory Fishes

There are numerous publications dealing with the use of fishes in a wide variety of research. Although it is not possible to give examples of all of these types of research, a few of the uses are mentioned here. Publications on the methods, techniques, and subjects of study by Klontz and Smith (1968) and Neuhaus and Halver (1969) were prompted by the increase in use of fishes in research.

Use of fishes in physiology is rapidly expanding because of the ease with which a great variety of factors can be evaluated in poikilothermic and aquatic animals. Pickford and Atz (1957) published a monograph on the study of the physiology of the pituitary in fishes, and the first two-volume textbook on physiology of fishes published in 1957 (Brown) has been expanded to six volumes (Hoar and Randall, 1969–1971).

Epidemic outbreaks of liver cancer in rainbow trout in America and Europe prompted coordinated research that resulted in a publication showing that the outbreaks were caused by aflatoxin, which is carcinogenic for this species of fish (Halver and Mitchell, 1967). Mawdesley-Thomas (1971) reviewed the use of fishes in a study of neoplastic diseases. More recently, the use of fishes in the teaching of etiology of bacterial diseases was recommended and described by Umbreit and Ordal (1972).

Fishes are also very suitable for genetic research as shown by Gordon (1953), Kallman (1970), and Kallman and Atz (1966). More specifically, the breeding of Koi, the Japanese colored carp, has been described by Axelrod (1973).

Most recently, increased pollution of water by industrial and domestic wastes has also prompted the use of fishes in large numbers because of their susceptibility to chemical and physical agents (Henderson and Tarzwell, 1957; Culley and Ferguson, 1969; Stephan and Mount, 1973).

V Sources

A. FRESHWATER FISHES

1. Fishes Obtained from Nature (Feral Fishes)

Vast numbers of fishes for research are collected from natural freshwater estuarine or marine habitats. These sources are acceptable if the uniformity and health conditions are not of paramount importance to the particular research. State laws govern seining in open waters. Therefore, permission from the fish and game authorities is necessary for collection of fishes. Great caution should be taken in seining because not only fishes but also plant material, mud, and other debris are netted. Gills may be covered with mud, and fishes may undergo severe stress in the suboptimal environment or otherwise be injured. During collection, scales or fins may be damaged or lost, opening portals of entry for pathogens. Damaged fishes usually die in a few days. Electrofishing results in stress that affects the sensitivity of fishes to toxicants (Lennon and Walker, 1964).

Styrofoam boxes or plastic bags are satisfactory containers for transporting fishes from the collection site to the laboratory. The Styrofoam material, in addition to being lightweight, is an excellent temperature insulator. Picnic jugs can be used to transport fishes for a few hours when temperature protection is important; however, their small size and the lack of gas exchange when they are tightly sealed may kill most fishes within 24 h. Transporting containers should be filled with water from the collecting site prior to seining, since the stream becomes roiled with debris during collection.

Fishes should be placed into the transporting container within 20 s upon removal from the water since fishes that are out of water longer may suffer brain damage due to lack of oxygen and/or temperature change. The transporting container is a poor environment since the quality of water in the container deteriorates rapidly and is subject to fouling with waste material from fishes. Water should be replaced with clean water taken from the same source at least 1 h after collection. The fishes may actively ram themselves against the sides and attempt to jump out, especially when suddenly disturbed. This can be partially avoided if a dark piece of cardboard is placed on the bottom of the transporting container, especially the white Styrofoam boxes. Since newly caught fishes generally do not eat, feeding is unnecessary and only contributes to fouling of the water.

A prolonged conditioning period is required for the fishes to adjust to captivity. Animals from the natural habitat should be held in isolation a sufficient time to determine their health status and to condition them to any changes in water, the general research environment, and the laboratory diet. Unfortunately, many species of fishes collected from nature cannot be economically maintained in captivity and few will reproduce. Therefore, fishes collected from nature should not be used if fishes reared in production facilities are available. However, almost all of the freshwater fishes common to the lakes and streams of North, South, and Central America may be induced to spawn, or they may be collected in the early spring when they are gravid and stripped for rearing in production facilities.

2. Hatcheries and Fish Farms

The culture of freshwater fishes is a well-established practice. Federal, state, and private fish hatcheries are producing fishes each year for restocking of natural waters, for human consumption, as bait for catching game fishes, and for domestic aquaria. Fish culture is a well-established branch of animal husbandry, and there are numerous sources of information on construction and operation of fish hatcheries or fish farms (Hickling, 1962; Snow, 1962; Tamura, 1966; Ghittino, 1969; Flickinger, 1971; Axelrod and Vorderwinkler, 1972; Bardach *et al.*, 1972; Heighway, 1973; Huet, 1973).

In hatcheries that serve as sources of laboratory fishes much effort is devoted to ensure standards of quality such as genetic history of the strain, age, parameters of growth rate and sexual maturity, nutritional state, health, and exposure to numerous environmental pollutants. The fishes obtained from natural habitats, however, may be more suitable for studies relating to the quality of the environment. If fishes are to be used in critical toxicity or chemical residue studies, some base-line data must be developed for each source of fishes for those parameters to be studied. A reference toxicant is often advisable to test fishes (Lennon and Walker, 1964).

The history of the care and maintenance of fishes obtained from hatcheries is usually known, and prior arrangements can be made to duplicate the hatchery procedures, techniques, diets, water quality, and other characteristics in the laboratory.

Even though hatchery fishes can be acquired readily, acquisition of large quantities must be planned to ensure an orderly flow of fishes to the laboratory. Research that utilizes juvenile fishes must be planned to coincide with normal periods of availability. An uninterrupted supply of juvenile or adult fishes may require one or more years of advanced scheduling with the supplier. Advanced scheduling assists the supplier in planning for the proper numbers and sizes of fishes required for research. Therefore, maintenance of a continuous supply of fishes to the laboratory requires communication between the researcher, culturist, and the supplying hatchery when scheduling is periodically revised.

Marking (1966) and Hunn *et al.* (1968) have described methods of handling and maintaining bioassay fishes, which, although still in the developmental stage, are useful for assay of reference toxicants. Most recently, Brauhn and Schoettger (in press) described culture methods currently in use at the Fish Pesticide Research Laboratory (FPRL) and the Fish Control Laboratory (FCL) for rainbow trout, channel catfish, fathead minnow, and bluegill.

Availability of fishes varies with the geographical area. For example, fingerling bluegills in the southern states can be obtained 60 days earlier in the year than those at more northerly latitudes. Periods during which fishes are available can be extended by selecting strains with different spawning times or by manipulating normal reproductive cycles. For example, twelve strains of rainbow trout were reported to have spawned at different times (Dollar and Katz, 1964). Thus, by appropriate scheduling of the requisitions, rainbow trout fingerlings might be obtained for testing throughout the year. Alternately, the periods of availability of embryological material can be extended by altering normal reproductive cycles (Hazard and Eddy, 1951; Carlson and Hale, 1972). For example, channel catfish normally spawn in early summer, and the fingerlings are available for research in the fall. Spawning can be altered by holding adults at relatively cool temperatures and then gradually increasing the temperature to stimulate the development of the ova (Brauhn, 1971). Largemouth bass (*Micropterus salmoides*) have produced spawn in the fall by use of similar techniques (Carlson, 1973).

Cost, the diet used at the hatchery, distance from the hatchery to the laboratory, water quality at both ends, availability, and genetic background of the stocks should be considered before acquisition of fishes from a hatchery. If fishes are needed on a routine basis, consideration should be given to a procurement contract in which stipulations of quality, dates, quantity needed, etc., are agreed upon and understood by both parties.

Reference strains of bluegills and channel catfish are propagated at two national fish hatcheries and are also used for pesticide research. Brood stock selection and maintenance of stocks has been practiced in North America for many years by both state and federal agencies. The Resources Agency of California, Department of Fish and Game's Rainbow Trout Brood Stock Program (Gall, 1972) is mentioned here as an example of fish availability from the state and federal hatcheries.

California has four primary brood stocks that are descendants of wild trout taken more than 70 years ago in California. These fishes have been subjected to an undetermined amount of selection to supply quality brood fish while at the same time an attempt has been made to improve the economics of the production of fingerling and yearling trout through genetic means. The genetics program in California is limited to a consideration of the size of eggs, number of eggs, percentage of egg mortality, size of fingerlings, and percentage of fingerling mortality. Various successes have been reported with the utilization of hormones to induce spawning in almost all fishes bred seasonally.

B. ESTUARINE AND MARINE FISHES

Estuarine and marine fishes do not readily spawn in captivity; however, approximately 50 or more species have been successfully spawned under controlled experimental conditions (Bardach *et al.*, 1972). The numbers of juveniles in these successful spawnings have been small; there has been no commercial success. One possible exception is the genus *Tilapia*, which thrives in a freshwater, estuarine, or marine environment.

C. TROPICAL FISHES

There is a wide variety of tropical fish species from which to choose in meeting the particular requirements of research. Culture of tropical fishes for aquarium trade is a well-established practice (Axelrod and Vorderwinkler, 1972). There are numerous farms in Florida and other tropical and subtropical parts of the world. At the present time there are only two sources of highly standardized tropical fishes for research in the United States. Homozygous platyfishes and swordtails developed by the New York Zoological Society are maintained and are available in small numbers at the Osborn Laboratories of Marine Sciences. The existence of a commercial source of a single genetically uniform clone of the gynogenetic *Poecilia formosa* propagated at Gossington Tropical Fisheries, Delray Beach, Florida, was announced by Agranoff *et al.* (1971).

VI Shipping Live Fishes

A. CONTAINERS

The shipping container should be light in weight, free from toxic substances, well insulated against heat or cold, and small enough to be handled by one person. The typical shipping box is made of 25-mm (1-in.) thick molded Styrofoam, with a tightly fitting lid, containing a tightly fitting polyethylene bag with a square bottom. The box and bag have inside dimensions of about 0.5 × 0.5 × 0.4 m (18 × 18 × 14 in.) at the bottom. The depth of the box is about 0.4 m (14 in.) and the plastic bag is about 0.6 m (24 in.) deep.

Only new bags should be used. The bag should be filled with 4 to 5 litres of conditioned water, after which the fish are added. The bag is then collapsed to remove as much air as possible, reinflated with pure oxygen, and sealed by twisting the top, folding the gathered plastic, and tying it with heavy rubber bands. Air can be used, but it furnishes only one fifth as much oxygen and is sufficient only for one fifth of the time or one fifth of the fishes as compared with pure oxygen (see Figure 1). The bag should completely fill the Styrofoam box, which is then inserted into a corrugated cardboard box (see Figure 2).

Labeling should include instructions such as LIVE FISH, DO NOT CHANGE WATER, KEEP AT ROOM TEMPERATURE, THIS SIDE UP, as well as the names, addresses, and telephone numbers of consigner and consignee.

B. ANESTHETICS AND OTHER DRUGS

It is customary to ship fishes in water in which they were raised. If necessary, to improve conditions during shipment, mild anesthesia and drugs can be added to suppress bacteria and external parasites. The most often used anesthetic is M.S. 222 (methanesulfonate salt of *m*-aminobenzoic acid ethyl ester) in a concentration of 1:45,000 (Bell, 1967). Oxytetracycline (Terramycin) 25–125 mg/litre (100–500 mg/gal) and methylene blue, 6 drops of a 5 percent solution (10 ppm), are used as suppressants of bacteria and external parasites. Acriflavin has also been used for bioassay fish. When the dye and oxytetracycline have thoroughly dissolved, the fishes can be introduced.

Many shippers use anesthetics to increase the ratio of number of fishes to unit volume of water. The long-range side effects of drugs and chemicals are not adequately known, and their use in shipping of fishes for research is not recommended. However, if usage of any drug or chemical is considered necessary, a declaration of the name of the compound and rate of application must be indicated on the label (Hunn *et al.*, 1968; McFarland and Klontz, 1969).

C. METHODS OF SHIPMENT

Although trucking is a reliable and economical method of transporting large numbers of young or adult fishes for short distances, air transport is probably the most practical for long distances and results in less stress. Air shipment is also the most practical means of transporting fish eggs over

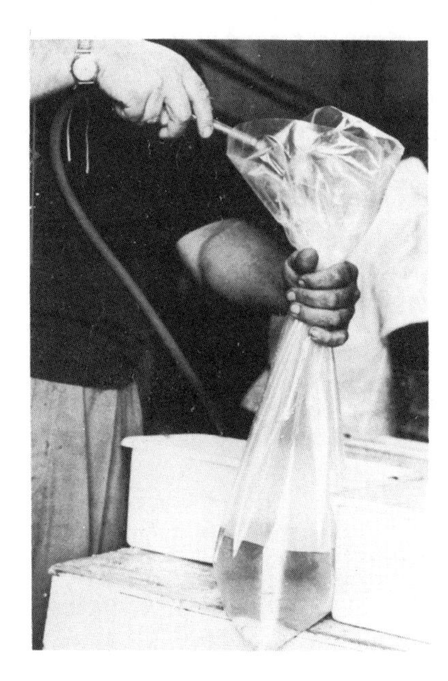

FIGURE 1 Pure oxygen being introduced into plastic shipping bag.

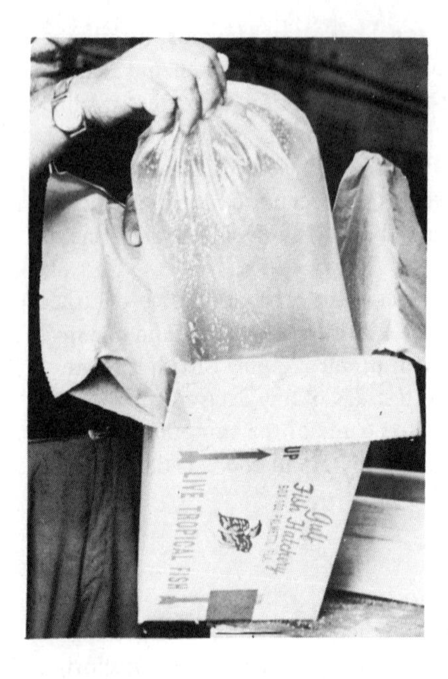

FIGURE 2 After the oxygen-inflated bag is sealed, it is inserted into the shipping box.

great distances. Shippers should notify the consignee of the itinerary of the shipment, giving specific information such as flight number and scheduled time of arrival. *Check all export and import requirements.* Pack the fishes immediately prior to departure. The primary hazards in transporting laboratory fishes are punctured bags, oxygen deficiency, excess of carbon dioxide and ammonia, and adverse temperature changes.

VII Isolation and Conditioning

All incoming fishes must be maintained in separate conditioning and isolation areas before being introduced into the laboratory. The fishes are preferably kept separate by species and within species by size and by environmental requirements. Each bay, aquarium, tank, or other fish container should be provided with separate color-coded dip nets, feeding and cleaning supplies, and other equipment needed during the holding period. Containers used to dispose of dead fishes, unused food, and decanted waste water should be conveniently located. No transfer of fishes, water, or any equipment between tanks is permissible.

Fishes obtained from nature should be suspected of being carriers of various parasites, bacteria, or viruses. Special attention should be given to the examination of the health of new shipments of fishes and to the diagnosis of the causes of any diseases. Any fishes showing signs of illness are not suitable for experimental purposes and should be destroyed. Physiological stress induced in wild fishes during netting, trapping, electric shocking, or transporting will generally render them unfit for research until they are conditioned to the laboratory environment (Lennon and Walker, 1964).

Fishes should be conditioned to temperature, water, diet, and the general environment. The length of the conditioning period should depend upon the nature of the experiment and be of sufficient time to determine that all animals are healthy and are feeding. Whenever practical, a conditioning period of at least 6 weeks is recommended.

Reconstituted fresh water (Lennon and Walker, 1964; Marking, 1966) and artificial seawater (Segedi and Kelley, 1964; Spotte, 1970, 1973) should be prepared in advance of receipt of the fish shipment so that the fish can be properly conditioned to the new environment. As a regular practice, regardless of the similarity of water characteristics, fishes must be conditioned to the water in the laboratory by slow dilution of the transport water. The temperature of the water used in transporting fishes should be as close as possible to that of the water in the laboratory; otherwise, it should be adjusted at the rate of 1 °C (2 °F) per hour.

VIII Care and Maintenance of Laboratory Fishes

A comprehensive treatise on the care and maintenance of all types of fishes used in the laboratory is beyond the scope of this publication. Cited references at the end of this report give supplemental material on the subject.

A number of existing facilities are good examples of the design and construction of holding and research facilities for aquatic animals (Hiatt, 1963; Spotte, 1970, 1973; Clark and Clark, 1971). Only the essential construction facilities for the holding and use of fishes are listed below.

A. INDOOR HOUSING FACILITIES

1. Design

The structure for housing of laboratory fishes must be capable of being easily modified to meet the needs of each new research project. The building must be constructed of steel, reinforced concrete, or comparable, structural, load-bearing surfaces that meet all local building codes. All tanks, aquaria, and water-holding reservoirs should be restricted to the ground floor unless the upper floors have been specifically constructed to carry high-density loads.

Large concrete tanks that are an integral part of the building are not recommended since they do not allow versatility and easy modification.

2. Structural Materials

All structural materials should be steel, concrete, or masonry products.

3. Walls

Since walls are exposed to high humidity, they should be impervious to moisture. The surface should be smooth and monolithic for ease of cleaning, preferably of ceramic tile, epoxy-painted concrete, or masonry; drywall construction is not recommended.

4. Ceilings

Ceilings should be waterproof and moldproof; they should contain recessed, waterproof lights (see Chapter VIII, Section A.8).

5. Floors

Floors should be made of concrete and have a rough surface to prevent slipping, but at the same time wash and clean easily. Floors should slope gently toward drains to prevent water accumulation.

Plastic tiles should not be used in areas where water is repeatedly splashed onto the floor. Floors should be concrete, painted with a waterproof paint; floor surface coverings should be monolithic and free from any cracks or spaces through which water may penetrate. In prefabricated buildings (built on a modular scheme) water leakage usually occurs at the junction of wall and floor or wherever there is an interruption in the floor to accommodate pipes and ventilation ducts; plastic sleeves do not assure the complete stoppage of the leak. There must be a solid waterproof bond between the floor and the walls. Floor drains, indispensable to accommodate water from leaky or broken tanks, should be accessible for easy cleaning. Floors should be kept dry.

6. Windows and Doors

Windows and doors should be made of corrosion-free materials. Windowless rooms may permit better temperature and vegetation control.

7. Ventilation, Temperature, and Humidity Control

Excessive ventilation may cause rapid evaporation of water from aquaria. Humidity may be controlled by means of ventilation or dehumidifiers to prevent accumulation of condensation on floors, walls, ceilings, and equipment. Temperature should be uniform to prevent condensation or excessive evaporation. Individual aquaria can be heated with thermostatically controlled heaters. These are available at most pet shops.

8. Illumination

Recessed fluorescent lamp illumination is desirable. Dimmer switches for incandescent lights only (rheostatic control of light intensity) and automatic illumination controls (to be set as needed in intensity and duration) are very desirable.

9. Power Supply

Standby power sources are essential and should be available for instantaneous hookup. Numerous electric outlets are necessary for water temperature control and operation of filters, aerators, and incidental equipment. All electrical equipment must be suitably grounded to prevent electric shock to fishes and humans.

10. Storage

Storage facilities are needed for aquaria, jars holding preserved fish, tools for handling fishes, and spare equipment. *Do not store chemicals together with feed or experimental animals.* Dry storage is needed for pelleted fish food, refrigerators for perishable materials, freezers for frozen foods, and experimental fishes preserved for future analysis.

11. Washing and Decontamination Facilities

Hot and cold water are needed for washing aquaria. Tanks are needed for disinfecting equipment in sanitizing agents, and steam sterilizers are needed for decontaminating the instruments, glassware, and culture media.

B. WATER QUALITY AND SUPPLY (GENERAL ASPECTS)

Adequate quantities of pure water must be available to supply the needs of the facility's carrying capacity for fishes. A complete chemical analysis should be made of the water for heavy metals, pesticides, dissolved salts, dissolved gases other than oxygen and nitrogen, and other toxicants, especially if water comes from other than domestic sources. Periodic checks may be necessary to monitor some contaminants of critical importance.

Laboratory water quality characteristics should be checked before designing new experiments to see whether some characteristics of the water quality are critical to the design (Brungs, 1973; McKim *et al.*, 1973).

The dissolved oxygen content in water is seldom more than 10 ppm and unless water is frequently exchanged or aerated the fishes may suffocate. Fishes tolerate only traces of metabolic waste products and will die unless the products are removed or changed to less toxic forms. The oxygen and carbon dioxide values are subject to considerable variability in water, particularly fresh water. The dissolved oxygen, carbon dioxide, and the ratio of free to ionized ammonia affect the utilization of dissolved oxygen by the fishes. This is demonstrated by the Bohr effect, which shows that where the carbon dioxide increases, additional oxygen is needed to maintain the same levels of oxygen in blood hemoglobin.

In slightly alkaline water, low in carbon dioxide, ammonia is present in the free form, which is more toxic than ionized ammonia. Users of fishes as experimental animals must be constantly aware of these changes which occur in an aquatic environment and must be able to prevent variation in parameters during experimentation.

The body surface and gills of fishes are exposed to all substances dissolved or suspended in water. For this reason, fishes are valuable animals for detecting and evaluating pollutants in water.

The great variety of water quality characteristics must be recognized and appropriately considered when fishes are changed from the water of one geographical location to that of another. There are more variables in surface waters than in subterranean sources. The extent of success in the care and maintenance of fishes depends upon the characteristics of the water quality.

Water with total hardness (as $CaCO_3$) between 150–250 mg/litre and temperature between 14–17 °C (57–63 °F) is optimum for long-term holding of native cold-water and warm-water freshwater species; temperatures of 23–29 °C (73–82 °F) are satisfactory for most tropical fishes. The required water flow depends upon the species being maintained, numbers and sizes of various species, and capacity of the facilities.

The characteristics of water quality that influence trout culture have been summarized (Leitritz, 1960; Davis, 1953). Since the temperature, pH, hardness, carbon dioxide, oxygen, and nitrogen of hatchery waters often differ significantly from those of the laboratory water, all fishes received in the laboratory must be conditioned (see Chapter VII). The rapid transfer of fishes to water differing radically in quality characteristics, such as pH and hardness, may result in severe shock or death (Holmes and Donaldson, 1969; Brauhn and Schoettger, in press). Whenever possible, fishes should be obtained from sources where the water quality closely simulates that of the laboratory. A water supply, such as a well or spring, with a constant flow of water with stable characteristics of quality, e.g., chemistry and temperature, is still considered superior to the recirculation system. If chlorinated water is used, chlorine can be removed with charcoal, sodium thiosulfate, or aeration. However, special care must be exercised to monitor the efficacy of the dechlorination system. A recirculation system for hatching freshwater fish eggs and rearing juveniles has been described (Guidice, 1966; Burrows and Combs, 1968).

Facilities for recirculation of the water are essential when a continuous supply of water of high quality is not available. The fish-holding capacity of recirculation systems is limited by the economics of filtering and pumping used water. Factors influencing water flow are discussed in detail in the following sections.

Water may be rendered useless for fish holding by leached zinc from galvanized pipes or copper from copper or brass pipes. To guard against possible metal ion toxicity problems, pipes that supply water to holding facilities should be made from black iron or high density polypropylene. Polyvinyl chloride, phenolic, or acrylic plastic pipes should not be used since these materials may contribute plasticizers, e.g., phtalate esters, to water.

1. Standardized Water

In experiments concerned with the biological effects of chemicals on fishes, chemical composition of water must be known and uniformly maintained. For this reason, so-called standardized reconstituted or reconditioned water is often used (Lennon and Walker, 1964). It is made from deionized or distilled water to which mineral salts are added in known quantities. Such salt mixtures must be made at the laboratory or obtained from commercial sources (Hoffman, 1884; Lyman and Fleming, 1940; Segedi and Kelley, 1964; Strickland and Parsons, 1968).

2. Water Supply for Aquaria

In the absence of the usual sources of satisfactory water, commercially bottled springwater may be used (see Chapter VIII, Section B.3). Soft water can be used without any treatment. Water can be dechlorinated as described above, or by letting it stand for several weeks. Storage basins should be of large size, for example 500–1,000 litres (125–250 gal), but may be of any shape, and should be made of glass or fiberglass.

Fiberglass ponds such as those sold in garden centers are excellent storage basins but may contain harmful plasticizers. They can be drilled so that water faucets can be attached. If the storage tanks are exposed to light, a layer of algae may develop along the walls, hastening the conditioning process. Nitrifying bacteria will grow along the sides and oxidize any ammonia which may be present (see Chapter VIII, Section B.3). Storage basins should be kept covered with glass to prevent evaporation and contamination. They may be placed either on the floor, in an elevated position on strong waterproof supports, or on a table. The elevated position has the advantage that water can be drained by gravity.

3. Conditioned Water and the Balanced Aquarium

An adequate supply of conditioned water is very important. It should be clear but may have a yellowish tinge. Conditioned water is "balanced" in the sense that it has an integrated microflora that oxidizes ammonia to much less harmful nitrates. The nitrifying bacteria, which need attachment to a supporting surface, multiply slowly; therefore, ammonia may accumulate in newly cleaned aquaria filled with unconditioned water. In a well-conditioned aquarium bacteria are found on the gravel, the plants, the aquarium sides, and the filter. These surfaces are unoccupied in newly set up and freshly cleaned tanks, although in those with conditioned water enough nitrifying bacteria are present in the water to serve as seed stock. It is useful to add conditioned water to the "new" water. The absence of colonies of nitrifying bacteria is partly responsible for the rapid deterioration of water in the collecting containers (see Chapter VI, Section A).

Small bits of food that are dropped into conditioned water will gradually disappear without causing the faintest trace of cloudiness in the tank. The food is slowly metabolized by bacteria into simple organic molecules. There is some evidence that in a well-established aquarium the presence of bacteriophages prevents a rapid increase in the populations of bacteria involved in protein breakdown. The same amount of food in "new" water will quickly increase bacteria and fungi that break down protein. The water appears cloudy from the great number of bacteria. The rapid depletion of oxygen under these conditions and the buildup of ammonia, plus the presence of other poisonous nitrogenous waste products and of carbon dioxide, cause unfavorable or lethal conditions for fishes. The ensuing change of pH in water may reduce the growth of beneficial nitrifying bacteria. It is best to introduce fishes gradually to a newly set up tank and to feed them sparingly or not at all for 3 or 4 days. An excellent account of the dynamic metabolic relationships that take place in a well-established, conditioned aquarium has been provided by Atz (1971).

Many scientists not familiar with maintaining fishes consistently underestimate the time required for the water to become fully conditioned. A period of a few days to 2 weeks is inadequate. Even if the fishes survive, the conditions will be suboptimal, and fishes may succumb under the stress of the experiment, leading to false results. In poorly conditioned water, fishes may hang motionless under the water surface or lie on the bottom with folded fins. If subsequently placed into well-conditioned water, they may recover.

Water quality may be tested by using spare fishes or a sensitive water plant. These fishes need little or no food and will at the same time condition the tank. Other indicators for water quality are *Daphnia* and snails, which are more sensitive to adverse conditions than are many tropical fishes.

Water must not be used for long-range experiments with aquarium fishes unless fishes have lived in it for a period of at least 2 months.

C. HOLDING FACILITIES FOR FISHES

In laboratories using large numbers of fishes, it is desirable to procure the fishes in advance and have holding facilities until they are needed. These facilities may also be used for isolation and preliminary observation on the suitability of the fishes for experimental use.

1. Indoor Holding Facilities

Indoor holding facilities are essential because fishes seldom can be obtained exactly when they are needed. The sizes and types of holding facilities are determined by the laboratory needs. Examples of holding facilities used at some testing laboratories have been described by Spotte (1970), Clark and Clark (1971), and Cullum and Justice (1973).

Exacting quantitative research should be done in artificial environments that are subject to a minimum of variation. A water-recirculating system plus a temperature control system such as that designed by Regier and Swallow (1968) or by Scott (1972a,b) will maintain small numbers of rainbow trout, channel catfish, fathead minnows, and bluegills at a desired temperature, Each of these four species requires time to acclimate to new surroundings. In most instances, a lengthy conditioning period will reduce the time that could be used in productive research. It is best to perform research in water with physical and chemical characteristics identical to those existing in holding facilities.

Prior to use, fishes should be maintained in facilities having a colored background similar to that of the experimental facilities. Channel catfish transferred from holding facilities with a blue-green background to research facilities with a white background are extremely sensitive to nearby movements and refuse to feed normally even though water quality and temperature in research facilities may be identical with those of holding facilities (Brauhn, 1971).

Fishes unaccustomed to holding facilities may respond to a sudden disturbance by jumping; therefore, holding containers should be covered with screening to prevent fishes from jumping out. If adequate covers are not provided, fishes found outside holding facilities should be discarded rather than returned to holding units. The skin abrasions caused by thrashing about on a hard surface will usually result in bacterial infection. Since channel catfish are negatively phototropic they acclimate faster to new surroundings if the tank is covered with black plastic. Brauhn (1971) used opaque plastic to induce channel catfish to spawn out of season.

Estimates of the proper number and weight of fishes an artifical environment can safely hold depend upon the species, dissolved oxygen concentration, feeding rate, temperature, and rate of water exchange in the holding unit. The metabolism and subsequent oxygen requirement of fishes held at constant temperature are directly associated with caloric intake (Fry, 1957).

There are many suggested methods of estimating the permissible weight of fishes per volume of water. A safety margin is always desirable in the loading ratio since these methods are only approximations. (For a detailed description of calculating loading factors for some species of fishes, see Appendix A.)

2. Outdoor Holding Facilities

Outdoor plastic pools and concrete and earthen ponds used for fish culture research have been evaluated by Shell (1966). Fishes in outdoor pools and ponds are exposed not only to naturally occurring light conditions and food chains but also to a variety of environmental contaminants. Plastic wading pools have a short useable life-span and may directly contribute to the chemical residues (plasticizers) and health hazards in research fishes. These factors may interfere with determining pesticide residues. Fiberglass pools are usually satisfactory.

D. HANDLING OF FISHES

Handling of fishes has often been referred to as an art. In reality, it is strict observation of feeding and cleaning schedules. Observation of any aberrant behavior is of utmost importance. *The reactions of fishes to efforts to maintain them and to their environment influence their value as experimental animals.* Behavior patterns of individual fish groups have been observed for 10 days prior to use as research animals (Hunn *et al.*, 1968); disease, malnutrition, and general conditions have been noted in the 10-day period (Lennon and Walker, 1964; Brauhn and Schoettger, in press). Other indicators of stress factors that subsequently influence experimental results include feeding response when food is offered, swimming posture or attitude in water, and unusual or erratic movements (Table 1). A change of behavior should alert investigators to look for stress-producing factors.

Observation of fishes intended for toxicological research should extend from acquisition through testing since many stress-producing factors may occur in a short time. Healthy fishes are usually very active during feeding.

1. Marking and Labeling of Fishes

There are many ways to mark fishes. Any method requiring removal of fishes from water and attaching or inserting a tag is traumatic and stressful. Following tagging, fishes should be observed until they no longer show signs of stress, unnatural behavior, or disease. The most frequently used methods of marking and labeling are:

- Numbered jaw tags (metal clips crimped into the maxilla)
- Numbered button tags inserted into fins or operculum
- Freeze branding
- Fin clipping
- Pigmentation patterns

2. Acclimatization

Newly arrived fishes, regardless of whether they come from another laboratory, a fish hatchery, or a wild population, have to be gradually acclima-

TABLE 1 Observations of Abnormal Fish Behavior and Some Common Causes for This Behavior[a]

Observation[b]	Possible Problem Area
1. Surfacing and swimming with mouth half emerged–a, b, c, d	a. insufficient oxygen; toxic chemicals; high NH_3
2. Scraping against sides of tanks–[c]b, d, f	b. parasitism of skin or gills
3. Erratic swimming–a, b, c	c. rapid temperature changes; parasitism of nervous system; virus disease
4. Crowding around water inflow–a, b	d. parasitism of intestine; bacterial disease
5. Distended abdomen and difficulty in swimming–d, e, f	e. diet consistency
6. Refusal to feed actively–a, c, d, e, f	f. nutrition deficiency
7. Listlessness, emaciation–b, c, d, e, f	

[a]Several observations may be made simultaneously and may be caused by a combination of two or more problem areas.
[b]Letters refer to possible problem areas.
[c]Normal fathead minnow courtship behavior includes coloration change and display near tank sides.
Source: adapted from Brauhn and Schoettger (in press).

tized to the laboratory environment. Fishes that have been overheated in shipment can be cooled gradually at 1 to 1.5 °C (2–3 °F) per hour. Caution should be exercised when the shipping container is opened as fishes may jump out. "Cooled" tropical fishes are thought to suffer less if the temperature is brought rapidly up to 21 °C (70 °F).

3. Social Factors

Fishes interact with each other not only within the aquarium but also through the glass between aquaria. Some species form hierarchial organizations that may be relatively stable with little obvious antagonistic behavior; others fight, are territorial, or form more or less well-defined aggregations. The type of interaction depends upon whether the fishes have been raised together from birth or were brought together as adults or subadults. How many fishes can be kept within a tank depends not only upon their sizes but also upon the size of the tank, aeration, filtration, water surface, and the type of social interaction (Lorenz, 1963).

4. Routine Care and Maintenance

An adequate assessment of stress-producing factors requires close adherence to a routine schedule of sanitation and feeding. Phillips (1970) suggests that small trout be fed small amounts of feed at hourly intervals in an 8-hour working day. Generally, during routine maintenance procedures the investigator has the greatest opportunity to observe behavior. For observation it is convenient to place small fishes into a small rectangular container that can easily be held in one hand. Round jars are unsuitable not only because they distort the view but also because it is difficult to net the fish in them.

Adherence to a routine maintenance schedule may condition fishes to a given set of circumstances, and any deviation from these conditioned responses must be considered when making a valid assessment of behavior. The care provided brood stock affects the success of the egg product and survival of the hatched fish (Davis, 1953).

5. Grading of Fishes

Fishes may be obtained as a population of mixed sizes. Specific research needs may require separation into uniformly sized groups. A mechanical fish-grader has been devised by Morton (1956) for fish hatchery use. Pyle (1966) later found that grading fishes into even-sized groups increased their growth rate. The spaced-bar type of grader (Pruginin and Shell, 1962) has been used at some laboratories.

Grading of fishes may cause severe shock and mortality when improperly executed. Only fine mesh nets made of a soft material should be used to transfer small numbers of fishes through a grader. Overloading the nets or graders can cause scale loss, skin puncture, or eye damage to some of the fishes. These injuries often are not readily visible and can become infected with pathogens, spread to healthy fishes, and result in loss of the experiment. For these reasons, grading of experimental fishes should be minimized and judiciously accomplished only when absolutely necessary The fishes should be observed closely for several days after grading to assess any possible damage.

E. AQUARIA FOR FRESHWATER FISHES

Aquaria come in various shapes and sizes from 5-500 litres (1-125 gal). The exchange of gas is facilitated by increasing the ratio of the surface to the volume of water. The most popular and useful aquaria are those with a stainless steel frame, slate bottom, and glass sides. Tanks with glass bottoms should be painted, preferably on the outside, since it is unnatural for fishes to receive light from below. In recent years more and more all-glass and plastic aquaria have appeared on the market. Their main advantage is light weight. Plastic aquaria have many disadvantages, including transparent bottoms and inclined and bulging sides that waste space; they are easily scratched and broken.

Aquaria must be covered to prevent (1) fishes from jumping out and contaminating adjacent tanks, (2) airborne dust from settling into the tank, (3) excessive loss of water through evaporation, and (4) chemical contamination.

Glass of 3 mm (0.1 in.) or 10 mm (0.5 in.) thickness is adequate as a cover for tanks with a surface area of up to 0.2 m^2 (2 ft^2); for larger surfaces, stronger glass is needed. Considerable time can be saved during feeding, especially when many tanks are in use, by cutting at one corner a triangular feeding hole approximately 38 × 38 mm (1.5 × 1.5 in.). The triangular piece can be attached to the cover with a piece of friction tape as a hinge. Edges should be polished to prevent cuts.

Large tanks can also be constructed with plywood sides coated with fiberglass or silicon material. Many coating materials have recently appeared on the market. Some epoxy cements are toxic; therefore, aquaria should be washed, preferably before the first use or when repaired.

The weight of tanks increases rapidly with volume. Tanks more than 200 litres (50 gal) should be placed in permanent position.

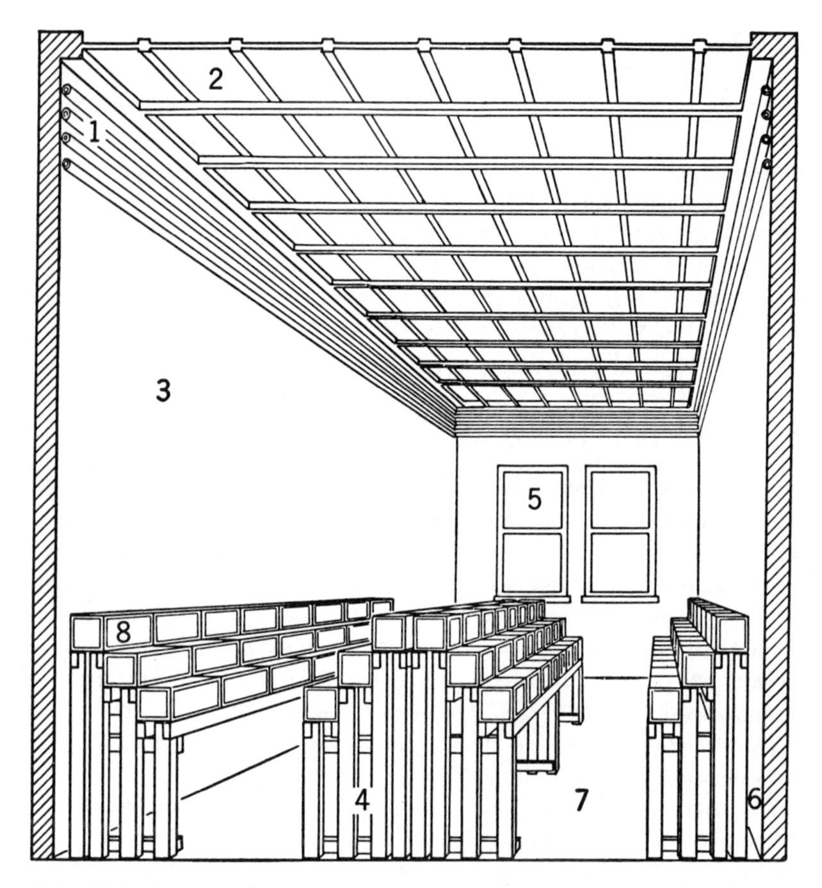

FIGURE 3 Figure representing interior view of the Genetics Laboratory of the New York Aquarium at the American Museum of Natural History. (*1*) Steam-heating coils offset the down draft of cold air from skylights. Another set of steam coils controlled by thermostats is located on floor near windows. The floor steam coils are not shown here. (*2*) Skylights are constructed of wired glass. Immediately beneath the skylights the laboratory has a system of shades, known as Ventilighters, for controlling the amount of illumination. The Ventilighters are not shown in this figure. (*3*) The sides of the laboratory are concrete. (*4*) The stands for the aquaria are constructed from 2 X 4-in. lumber; for detail see Figure 4. The central racks are placed back to back. (*5*) Windows for ventilation are usually opaqued on the inside to exclude excess light. (*6*) The racks near the walls are placed 4 in. away from the wall to avoid the colder concrete. (*7*) The aisles between racks are 36 in. wide. The floor is of concrete and has a drain built into it. The laboratory has facilities for artificial illumination, compressed air, and a sink with running hot and cold water. (*8*) The aquaria are arranged in three tiers. (Drawn by Donn Eric Rosen.)

1. Handling of Aquaria

Aquaria must be moved when empty. They should be supported with two hands only from below, along the metal frame. When carrying aquaria, pressure should not be applied to the bottom. Aquaria are best stored filled with water to prevent the aquarium cement from drying, which may result in leakage. All-glass tanks may be stored dry.

2. Stands for Aquaria

The tanks should be placed on specially designed stands composed of wood or noncorroding materials (Figures 3 and 4). The wood should be varnished. Pine is acceptable provided it is free of knots. As shown in Figure 4 the entire stand is made of 50 X 102-mm (2 X 4-in.) lumber held together by corrosion-resistant nuts and bolts. The entire structure can be taken down and reassembled elsewhere within a few minutes. The racks of tanks and aquaria should be placed back to back away from the walls. The design can easily be modified for tanks of various widths. The tanks on each stand need not be of the same width. The stands can be of any length, but most convenient are those 2.75 m (9 ft) long with a center support at 1.37 m (4.5 ft). Longer stands may prove to be unwieldy should they be needed in a small room. The space below the tank rows, left empty in Figure 3, can be used for *Daphnia* culture, water storage basins, or spare aquaria.

The tanks should be supported only by the runners along the edge. This is desirable since pressure (support) must only be applied to the metal frame of the aquarium, never to the slate bottom. The tanks must rest on a flat surface (wooden board), or a system of runners to permit water and dirt to fall to the floor where it can be flushed away by a hose or swept up. Rows of tanks must never be above each other. Instead, the rows should be staggered as shown. This system also prevents dirt from falling into the aquaria below.

The aisle between the stands should be a minimum of 0.75 m (30 in.) wide. If metal stands are used, they must be corrosion-proof. In the center of the room the stands are arranged back to back. The three rows of each stand shown in Figure 3 are designed for aquaria 0.23 m (9 in.) wide, and the design can easily be modified to accommodate aquaria of different widths in each row. Care should be taken that the height of the tanks is such that those in the front do not extend above the bottom of those in the following row.

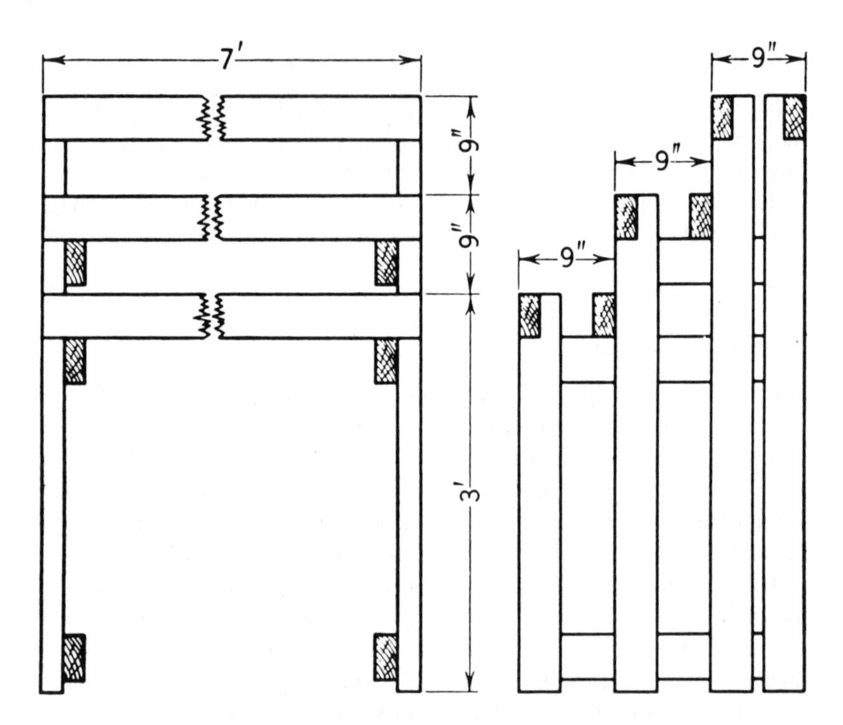

FIGURE 4 Figure of stand for a battery of 15 aquaria, each of which measures 16 X 9 X 9 in. To the right, the end view shows the arrangement of the 2 X 4-in. uprights and the ends of the 2 X 4-in. planks that serve as runners upon which the aquaria rest. To the left, the front view shows the arrangement of the runners; the narrow 2-in. edge is placed uppermost. (Drawn by Donn Eric Rosen.)

3. Effects of Illumination

For all experiments where light intensity and photoperiod may be of critical importance, the fishes must be maintained under strictly controlled conditions. Although fluorescent lights provide a uniform source of light and give off little heat, some emit ultraviolet radiation that may be harmful (Perlmutter and White, 1962). If a programmed light sequence is desired, the apparatus by Wickham *et al.* (1971) may be incorporated into indoor lighting systems.

Aquaria should not be set up near windows because the strongest light should not come in through the sides of aquaria, but rather from above. Where light comes in through the sides, fishes do not swim in their normal attitude, but will incline their bodies at about a 30° angle towards the side (light source). Their conflict between positioning themselves according to gravity or to light source could create stress.

Excessive light intensity and especially direct sunlight causes a rapid growth of blue-green algae that will cover in broad sheets all surfaces (including the desirable aquarium plants) inside the aquarium. Also, the water may turn green from the presence of a bloom of unicellular green algae. This condition can be harmful to the fishes, by production of nitrogenous products and elevation of pH, and also may cause oxygen depletions upon decay. Both conditions can be controlled by reducing the amount of light; the blue-green algae will separate from the surface and can be netted out. Green water is easily cleared by first removing the fishes and then adding a few small crustaceans such as *Daphnia.* The latter will multiply rapidly and eliminate the green algae by feeding on it. Strong light is harmful to some aquarium plants such as *Cryptocoryne*, which grows in rather dim light on the bottom of jungle streams.

Sunlight may also heat the water in the experimental unit. Strong illumination by sunlight of a heavily planted aquarium causes a temporary excess of oxygen and depletion of carbon dioxide due to an increased level of photosynthesis. Rapid utilization of carbon dioxide from carbonates may lead to alkalinization of water to such an extent that it may be toxic or even lethal to fishes. As long as the temperature is kept within the range that is "natural" for the species, many tropical fishes will stay in breeding condition whether the photoperiod is constant, increasing, or decreasing. For species that live in high latitudes, increasing temperatures and light periods—often in a particular sequence—are required. After the breeding season the individual often has a refractory period.

Light, particularly the daylight fluorescent type, may be lethal to eggs or small fishes. Be sure that the illumination used is harmless (Perlmutter, 1951; Lyubitskaya, 1956; Eisler, 1958; Schnarevich, 1960; Lyubitskaya and Dorofeeva, 1961; and Steucke *et al.*, 1968). Dimmer switches and photoperiod adjustment may be desirable for many native species. At the Genetics Laboratory associated with the New York Zoological Society poeciliid fishes (*Xiphophorus*) are routinely bred throughout the year in large rooms with skylights. During the winter additional incandescent light is provided for 5 h.

There is some evidence that fishes will breed better under natural light conditions than under artificial illumination. Therefore, breeding facilities for tropical fishes should be constructed with a skylight, and the walls should be windowless. Louvers should be installed in the skylight and baffles installed above them, running at right angles to the louvers, to intercept the direct rays of the sun.

4. Effects of Color

Fishes must have a healthy background coloration. That which is "healthy" and "normal" has to be learned gradually by the experimenter. Fishes that have a dark appearance or are too light are obviously stressed.

Certain species exhibit pigmentary polymorphism. Some polymorphisms are considered "natural" because the different color variants are found in natural populations, e.g., the guppy and the platyfish. Other polymorphisms are said to be "artificial" because they arose under domestication. Good examples of the latter are the xanthic and albino forms of many species.

Coloration is influenced by the diet, breeding season, social order, salinity of the water, the light conditions to which the tank is exposed, and the emotional state of the fish.

5. Filling Aquaria

Tanks should be covered and kept filled to a specified level. Evaporation may cause a mineral deposit in the form of a whitish ring around the aquarium at the water surface, depending upon the hardness and amount of water lost. In rooms with forced air circulation, aquaria in the direct path of the air flow will lose considerably more water than those in other locations. Under such conditions aquaria may lose up to 11 percent of their water in a week. In laboratories with many aquaria, where tank maintenance consumes a certain amount of time, it may be more economical to discard such tanks or to place them in a location without drafts. Excessive loss of water and variable levels cause a rapid increase in the mineral content of the water. Loss of water is also unacceptable in maintenance of fishes where osmolarity of water must be maintained.

6. Water Removal

One fourth of the conditioned aquarium water can be withdrawn from active tanks as needed and replaced. Do not tap any tank more frequently than once per month. On the outside of the tank the water removal date and the amount taken should be indicated with a wax pencil. Water should be replenished from the storage tanks. It is best not to remove water from a tank with juvenile fishes.

Waste water from experimental fishery research facilities cannot be emptied directly into any waterway without prior treatment. This water must be filtered and treated to remove not only escaped fishes but also their eggs and parasites and other pathogens. Decontamination is essential for the water used in disease studies. *Consult all relevant state and federal laws regarding disposal of refuse water–especially if it contains drugs or other chemicals used in maintenance of health.*

7. Controlling Water Temperature

If water of different temperatures is needed, storage tanks with warm and cool water are desirable. Water at any desired temperature can be obtained by delivery from the storage tanks with temperature-controlled mixing valves (Scott, 1972a).

Freshwater and marine fishes can be grouped, according to their optimum requirement, into the water temperature ranges given in Table 2.

Except for special experiments, the water temperature must be maintained within these ranges of temperature for each group, and the thermostats should maintain it within 1-2 °C. The rate of change in water temperature is critical and should be varied slowly, 1-2 °C per hour, or the fishes may be subjected to thermal shock. Slow temperature change is also required for acclimatization of fishes to those test conditions outside the above ranges.

Auxiliary equipment must be maintained for the control of water temperature. The choice of the method depends upon the type of control desired, the size and design of the facilities, and the number of tanks. If few tanks are involved, this is best accomplished with thermostatically controlled heaters in each aquarium. However, this solution is totally impractical in places where several dozens or hundreds of tanks are in use. In such large facilities an independent alarm system set for the maximum and minimum acceptable temperatures must be installed. Duplicate heating systems (i.e., more than one boiler, and independent pipes and ducts to radiators and vents) are the best insurance against a disastrous heating failure during the winter months. Because of the thermal capacity of water even a heating failure of up to 12 h on a cold day will lead only to a temp-

TABLE 2 Water Temperature Ranges of Freshwater and Marine Fishes

Trout and similar cold-water species	10–15 °C (50–59 °F)
Minnows, bluegills, catfish, goldfish, and closely related species	20–25 °C (68–77 °F)
Warm-water tropical, freshwater, and marine species	22–30 °C (70–85 °F)
Marine (depending on species)	6–30 °C (43–85 °F)

erature drop of 4–5 °C of the aquarium water inside a well-constructed building. The larger the tank, the less the temperature variation.

Aquaria used for experiments in which temperature is critical should be placed in specially designed temperature chambers that are available commercially. It is much easier to keep an aquarium above rather than below ambient room temperature. Aquaria can be held at lower than room temperature by using refrigerator units or cold rooms.

8. Aeration

A dependable source of compressed air is needed for aeration purposes and for most filters. Compressed air can be generated within the fish room by a variety of portable, electric-powered air pumps, each of which can serve several tanks. The use of portable pumps, however, can lead to clutter and the loss of valuable space. A much more efficient system, especially when there is more than one large fish room in the building, is a central source of compressed air. Air should be compressed by rotary pumps or "blowers," or oilless pumps. Although oil filters are available, never use oil-lubricated piston pumps since oil will eventually find its way into the lines (Scott, 1972b).

Air outlets should be provided near aquaria stands. A plastic tube can then extend along the front of a runner on which the aquarium rests. If more than one line is needed from the same outlet, they should branch close to the source.

The volume and pressure of air delivered to the most distant outlet should be checked periodically with all outlets in the open position. The central system can be more easily monitored than can the several individual units; a pressure gauge and periodic volume air-flow check usually are sufficient. A standby unit is essential.

9. Filtration of Water

The function of a filtration system is to remove excess food and feces and to eliminate some soluble products of metabolism, such as ammonia, by means of nitrifying bacteria growing on filters. The additional functions of filtration are aeration and circulation of water, removal of carbon dioxide, and replenishing the dissolved oxygen. The water current through the filter is created by an airlift (a source of compressed air is necessary) or by a water pump. The water passes through a layer of activated charcoal, nylon wool, or gravel, which traps particulate matter.

Some filters are placed inside the aquarium; others hang on the outside.

A "corner" filter is placed into the rear corner of an aquarium and air is bubbled through it. Corner filters are inexpensive and easily cleaned, and save space since the tanks can be placed close together. They are efficient in tanks up to 57-liter (15-gal) capacity. One disadvantage is that some fishes may hide behind them and go underneath them. A small triangular piece should be cut from the rear corner of the tank cover to accommodate the air line.

"Outside" filters, easily cleaned and more efficient than inside ones, are narrow plastic boxes, roughly 0.23 X 0.05 X 0.18 m (9 X 2 X 7 in.), that hang on the outside of an aquarium. Water is drawn into the filter through an inverted U-shaped tube. A common problem is that the arm of the U-shaped tube that extends into the aquarium is too short. Consequently, any loss of water in the tanks will leave the mouth of the filter arm above the water level and the flow of water will stop.

Under-gravel filters are esthetically more desirable and rarely have to be cleaned, but when they are cleaned, all plants, gravel, and fishes must be removed.

There is no doubt that filters increase the carrying capacity of an aquarium and are necessary for certain species with high oxygen requirements and high sensitivity to ammonia. However, a large number of species can do very well without either aeration or filtration, which take up valuable space and require time for servicing.

Larger production units utilizing biological filters have been designed for production facilities used in rearing native species (Burrows and Combs, 1968).

10. Sand and Gravel

If the purpose of an aquarium is to provide an acceptable habitat for long-range experiments or for breeding freshwater fishes, the presence of sand and gravel may be essential. Gravel of No. 2 commercial grade should be placed into the tank. The layer of gravel should be highest toward the rear wall of the aquarium [about 12-25 mm (0.5-1 in.)], depending upon the aquarium size, and then slope gently downward toward the front. The anterior one third of the aquarium should be free of gravel. Feces and other solid matter will gradually fall down the slope and accumulate as large flocculent mass (mulm) in front of the gravel where it can be siphoned off.

Coarse gravel is unsatisfactory because food particles that may become lodged beneath it will remain uneaten and eventually rot, and plant roots will die because of lack of oxygen. New gravel must be thoroughly washed to remove all fine dust particles or any soluble toxic material.

11. Plants

Plants, besides being decorative in an aquarium, supply hiding places for fishes, especially those species that are partial to weedy waters. Unless the aquarium is densely planted, the vegetation has only a negligible effect on the oxygen and carbon dioxide concentration in the water but may affect nitrogen compounds and pH. Ammonia and excess alkalinity can be a problem when plant density is too great. Buffers may be required to correct the situation. Not more than three or four plants should be placed into a small tank so that the fishes can be easily observed. The roots should be placed in the gravel near the rear wall, one plant in the center, the others halfway toward each rear corner. Plants in the corner interfere with the netting of fishes.

Plants are not useful in aquaria occupied by fishes that dig in the gravel. As a replacement for plants, flower pots laid on their sides, plastic pipe sections, or rocks may provide hiding spots.

Hardy plants with good root systems and leaves are *Echinodorus intermedius* and various species of *Cryptocoryne.* Good floating plants are the *Eichhornia*, *Lemna*, *Nitella*, *Myriophyllum*, and *Riccia*.

12. Labeling of Aquaria

A permanent label that utilizes a water-activated adhesive should be affixed to each tank after cleaning. Most labels will stick to glass but a large variety of labels will not adhere to plastic aquaria. The label should always be placed in the same corner with the lower margin of the label above the water surface so that fishes cannot hide behind the label. A duplicate of the label should be entered into the permanent record book of the laboratory.

13. Cleaning of Tanks and Aquaria

Probably more time is spent on cleaning tanks than on any other task. If this is not done properly, tanks may be damaged or the water may become cloudy after cleaning. The following method enables tanks to be cleaned in 15 min or less:

(a) All fishes must be removed from the tank; the plants, if present, are uprooted; and the person in charge of fishes or experiments clearly indicates on the label that the aquarium is to be cleaned. This is a safety measure to prevent fishes from being accidentally discarded. The tank is left standing empty for at least 24 h to allow sufficient time for the detritus or muck to settle before cleaning. Green water caused by abundance of algae can be cleared up with *Daphnia* (see Chapter X, Section C).

(b) Sufficient numbers of suitably sized pails should be placed on the floor in front of the aquarium so that the siphoning process can run without interruption. The opening of the siphon should be 50–100 mm (2–4 in.) above the floor and muck of the aquarium. Siphoning should be stopped when 25–50 mm (1–2 in.) of water is left in the tank. A good siphon is a fairly stiff garden hose with a 13-mm (0.5-in.) inside diameter and is about 2 m (6.5 ft) in length for tanks 1–2 m (3–6 ft) above ground. Soft plastic tubing collapses where it is bent over the edge of the aquarium, stopping water flow.

(c) The tank should be taken to the sink. It should be supported, pulled, or carried only by its metal edge; however, a full aquarium should never be carried. A convenient sink measures 0.92 m (36 in.) long, 0.5 m (20 in.) wide, and 0.3 m (12 in.) deep with a wide drain shelf on one or both sides for the tank to rest on. The faucet should have a swivel head, be 0.13–0.20 m (5–8 in.) above the upper level of the sink, and extend about 2/3 the distance toward the front.

(d) Plants uprooted the day before are placed in a pitcher with storage water. (Remove runners and side shoots; leave only nicely sized plants.)

(e) Gravel and dirty water should be poured from the tank into plastic buckets. A thin sheet of metal (stainless steel or aluminum), plastic, or a spatula about 50–150 mm (2 X 6 in.) can be used to scoop up remaining gravel. The bucket with gravel in it should be placed under the faucet, filled with hot water, stirred to remove dirt, rinsed in cold water, and removed from the sink.

(f) The tank should be washed in lukewarm water either on the drain shelf or in the sink. The glass sides and bottom should be cleaned with a fine steel wool pad soaked in a laboratory-approved detergent such as Alconox and a sponge to get rid of algae and dirt. The outside of the aquarium should be rinsed. When thoroughly soaked, labels will come off without effort. The tank should then be rinsed thoroughly with lukewarm water.

(g) The steel wool should be returned to the solution of Alconox to prevent rusting.

(h) Clean gravel and plants should be placed into the tank and the tank returned to its place.

(i) Gravel should be positioned carefully along the rear of the tank except for a 50-mm (2-in.) space in either corner.

(j) The label should be placed in the upper-right-hand corner of the tank.

(k) The tank cover should be placed inside at a slant so that its edge rests against the bottom of the front glass while being supported further back by the upper rear edge of the aquarium.

(l) The water stored in the buckets should be poured on the glass cover so that the gravel is not disturbed. Enough water should be added to fill the aquarium to the lower edge of the label.

(m) The cover should be placed on the tank so that the feeding hole is over one rear corner.

(n) The hose, buckets, and other cleaning equipment should be returned to their proper places after disinfecting.

(o) The sink should be cleaned.

14. Nets

Nets come in various sizes and shapes and are adaptable to different uses. The frame of the net should have sharp corners since it is virtually impossible to catch fishes in aquaria corners with rounded nets. The net should be composed of thin, soft, very pliable material and should have very thin seams. Nets with very fine mesh are useful for *Daphnia* and brine shrimp nauplii but less suitable for capturing fishes since water passes through slowly.

Since diseases and pathogens can be transferred by nets, it is recommended that sanitizing agents and sanitary techniques for nets be used. *Never use a net in more than one aquarium or tank without sanitization.* Nets treated with chemical disinfectants must be rinsed in clean water. Carry-over of disinfectants on nets may be toxic or lethal.

F. LOADING FACTOR

The holding capacity of a tank is the number and weight of fishes in relation to the volume of water that can be maintained without any ill effects under specific conditions of temperature, aeration, and feeding. The holding capacity also depends on the social behavior of fishes. Incompatibility causes severe stress, and incompatible fishes should not be kept together except in social behavior experiments. Loading factors for some species of fishes frequently used in research and assay are given in Appendix A (Lennon and Walker, 1964).

G. DISPOSAL OF FISHES AND WATER

Live or dead fishes and water containing toxic chemicals or microorganisms pathogenic to humans and animals should never be released to nearby streams or any other open bodies of water; nor should they be discarded into the sewer system without decontamination or neutralization.

There are various local, state, and federal regulations governing the treatment of effluents from research facilities, fish hatcheries, and animal holding facilities. These regulations are often changed and are too diversified to be included here. All persons using fishes in research should become acquainted with and comply with such regulations.

There are new and specific regulations in the Refuse Act enforced by the Environmental Protection Agency (PL-92-500, Federal Pollution Control Act, Amendment of 1972) governing the release of wastewater from fish culture and holding facilities to open waters. There is also the proposed Hazardous Waste Management Act (HR-4873) dealing with disposal of contaminated water (Hinshaw, 1973).

Proper containers and refrigerated facilities must be provided for the storage of dead fishes until they are incinerated or disposed of in any other approved manner. Where facilities are available, contaminated waters may be disposed of in a dry well and decontaminated by addition of 100–200 ppm of chlorine. If the fishes are still alive, they should be killed by immersion in water containing an anesthetic. Dead fishes should be incinerated or, if appropriate facilities are not available, should be buried deep with the addition of lime or hypochlorite. Use of incineration for disposal of fishes is preferred for those subjected to drugs or chemicals, as required by the Food, Drug, and Cosmetic Act (Sec. 107, PL-87-781, Drug Amendment Act, October 10, 1972).

Certain municipalities require that wastewater from chemical laboratories pass through white marble chips to neutralize acids before entering the public wastewater systems. It should be recognized that the soft detritus (mulm) at the bottom of fish tanks, when flushed down the drain, will eventually clog the spaces between the marble chips. The detritus from several dozen tanks is not sufficient to cause any serious stoppage, but if the laboratory has several hundred fish tanks, a water outlet separate from that of chemistry laboratories should be provided.

H. DISINFECTION AND STERILIZATION

Certain sanitizing agents, such as Chlorox, Novalson, and other chlorine-releasing compounds, may be used for sterilization or disinfection of equipment and facilities. Quarternary ammonium disinfectants, such as Roccal and Hyamine compounds, are also effective, colorless, and odorless. They are toxic to fishes and have to be washed out of tanks, nets, and other equipment before these come into contact with fishes. All disinfectants should be used as indicated on labels.

Sterilization can be accomplished with very hot water, steam with or without pressure, or hot air provided that equipment is not damaged by heat. Ultraviolet irradiation of water is also effective if the water is not turbid and the exposure time and emission are adequate.

I. SPECIAL FACILITIES FOR THE HANDLING OF MARINE EXPERIMENTAL FISHES

The study of marine animal diseases depends upon the availability of a system for maintaining specimens under controlled and defined conditions. Keeping marine fishes in aquaria or in closed, artificial, recirculated seawater systems is not a difficult procedure, provided that temperature, osmotic pressure, alkalinity, and chemical composition of marine water meet the critical requirements of marine fishes kept in the laboratory (Ellender *et al.*, 1973).

Large-scale fish production and maintenance facilities have been investigated, and excellent descriptions of these systems are readily available (Klontz and Smith, 1968; Spotte, 1970, 1973; Clark and Clark, 1971; May, 1971; Shelbourne, 1971; Bardach *et al.*, 1972). The facilities described below are for the smaller investigators and are only for the maintenance of marine fishes in controlled environmental conditions.

1. Facilities

The physical facilities needed will depend upon the size and design of the proposed project. For moderate-scale aquaria and laboratory operations adjacent to marine waters, the collection of papers by Clark and Clark (1971) is sufficiently comprehensive to provide adequate guidelines.

For the smaller operator and for investigators far removed from coastal areas, the physical facilities and construction design are very important. All building materials should be resistant to the corrosive damages of condensations and drippings of high-salinity waters needed to support the marine fishes. Walls, floors, and ceilings should be of salt-resistant materials that are easily washed, to facilitate removal of excess accumulations of salt deposits.

Room enclosures should be adequately heated and ventilated. Full temperature control is ideal but not always available. Room light fixtures should be moisture resistant and explosion proof, such as those used in refrigerated facilities. Where small numbers of fish are maintained in aquaria or tanks, the recirculated aquarium water temperature may be controlled by an immersion-type heater or refrigerant. In larger facilities centrally located heating and cooling equipment may be more economical and convenient. All electrical control panels and outlets should be of the marine outdoor type to prevent shorting and arcing from saltwater condensates.

All plumbing fixtures should be constructed of chemically inert materials and exposed for easy service. In general, all metals, including stainless steel, should be avoided; valves, joints, and other critical areas should be constructed of Nylon or Teflon. Nonplasticized polyvinyl chloride or Pyrex glass piping should be used throughout the system.

Wash water, discarded aquarium or tank water, and cleaning fluids should be emptied into a drain constructed of noncorrosive materials. All water should be treated prior to being discarded.

The water supply should come from an uncontaminated source. If pumped from estuaries or from the ocean, it should be filtered or run through settling and aerating basins before use. For the smaller operation, tap water should be deionized and/or distilled prior to addition of synthetic seawater-salt mixtures.

2. Aquaria

Some aquaria cements may deteriorate and release toxic breakdown-products into the saltwater. Silicone cement, which is inert, strong, and nontoxic may be used to repair and seal glass or plastic aquaria.

Correct handling of the water supply, either fresh- or seawater, is essential. Water collected from the sea may be filtered, stabilized by storing in the dark for 1 or 2 weeks, or decontaminated with ultraviolet light.

Several commercial, synthetic sea-salt mixtures are available for use in the laboratory. These synthetic salts are dissolved in deionized or distilled water as directed by the manufacturer. Synthetic seawater should be aged 20–30 days before use and held at a salinity of 30 parts per thousand (ppt). Prior to use the salinity is adjusted to the desired concentration with a hydrometer (specific gravity), an osmometer (freezing-point depression), or a refractometer (American Optical Company, Buffalo, New York). Refractometry is the more accurate and versatile of the three methods. Comparative salinity measurements using each of the three methods are shown in Figure 5.

The pH of the water should duplicate the range of pH found in nature (pH 7.5–8.4) for optimum results. Nitrate buildup must be avoided as this may lead to acidification of the water system.

The oxygen concentration must be maintained within close tolerance, and it should never be less than 2–3 ppm, depending upon the species of fish. Water should be aerated continuously. The air supply system must be oilless or be equipped with a water trap to remove impurities that may injure the fishes.

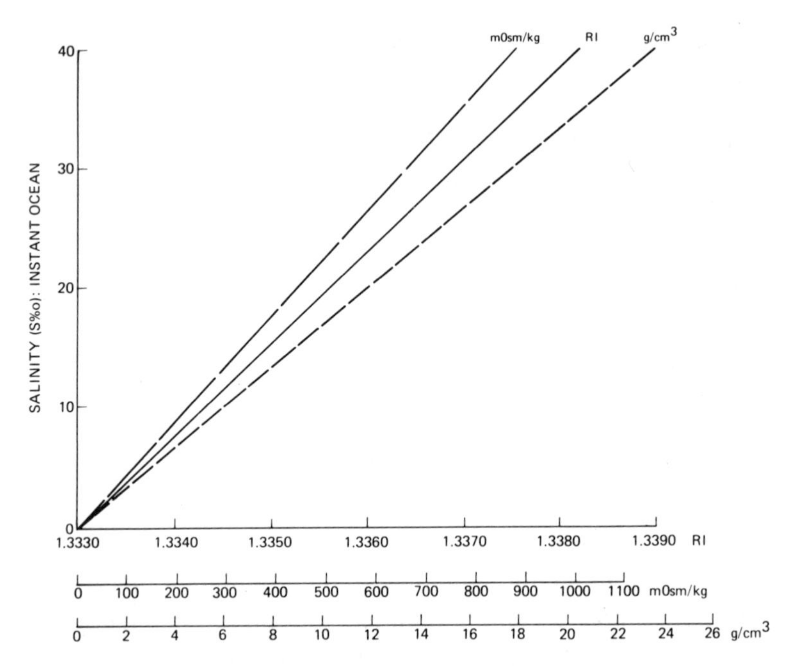

FIGURE 5 Standard curves for the determination of salinity (S%o) in artificial seawater (instant ocean) by the methods of specific gravity (g/cm^3), osmolarity (mOsm/kg; milliosmosis NaCl per kilogram) and refractometry (RI; Refractive Index).

Recirculation and filtration of the water are also essential for removal of accumulated waste products, toxic materials, and unused food. A Jet Power Filter (Turtox Corporation, Chicago, Illinois) or equivalent unit layered according to the procedure of Parisot (1967) is recommended for the small-scale unit. This mechanical filter unit is composed of (top to bottom) crushed oyster shells, graded sand, glass wool, activated carbon, gravel, and a fiberglass mat. This type of system provides a buffering capacity, stabilizes pH, and removes debris. The filter unit must be replaced at regular intervals (2–4 weeks) for optimum results. The rate of water flow should be adjusted to approximately 10 percent of the tank volume per hour.

Ammonia toxicity in the closed, recirculated seawater system is a serious hazard. It is produced by decomposition of uneaten foods, dead fishes, plants, and other organic matter, and is the principal excretory product (urine and feces) of fishes. Excess levels of ammonia may be controlled by (1) dilution (replacing part of the aquarium water at regular intervals—25 percent per mo), and (2) introduction of nitrifying bacteria in the filter bed. A simple method for determining ammonia-nitrogen levels in artificial seawater has been described by Ellender *et al.* (1971).

Many types of aquaria are suitable for maintenance of marine fishes. The majority are adaptations of or are identical to the systems used for freshwater tropical fish studies. A workable small-scale system is shown in Figure 6.

J. TROPICAL (ORNAMENTAL) FISHES

Many widely available species and groups of ornamental fishes have been selected for study since they are relatively easy to maintain and reproduce readily. They are suitable for all types of research or observations. They vary greatly in the manner in which they reproduce and take care of their offspring. On this basis they can be divided into several groups, each having

FIGURE 6 The rack-type unit is commercially available from Research Equipment Company, Inc., Bryan, Texas, in four- or eight-unit racks.

FIGURE 7 A female guppy delivering live young.

different advantages as laboratory animals. Their maintenance in the laboratory is briefly described below.

Tropical fishes rear their young in four manners: as *livebearers*, fishes whose young are delivered alive, either viviparously or ovoviviparously; as *egglayers*, fishes that either scatter their eggs, deposit them on a solid base, or bury them in the sand; as *mouthbrooders*, usually cichlids and some anabantoids, that lay their eggs on the substrate and then carry them in their mouths until the young are large enough to fend for themselves; and as *bubblenest builders*, fishes whose eggs are deposited in a foam nest.

1. Livebearers

Among the livebearers most readily available and best known are those belonging to the family Poeciliidae (Jacobs, 1971; see Figures 7 and 8). Usually, poeciliids are the most easily maintained and easily bred fishes. One male can fertilize many females and the females can store the sperm and fertilize successive broods without subsequent mating. Depending upon the size and number of fishes, most livebearers can be housed in 4–20-litre (1–5-gal) aquaria. Of the livebearers suitable for laboratory purposes, the platyfish, the swordtail, and the guppy are the most readily available. Other species of the genera *Xiphophorus* and *Poecilia* (*Mollienésia, Limia*), as well as the smaller mosquitofish, should be considered.

A stocking rate for good maintenance is expressed in length of fish per volume of water. As a general guide, with efficient management, the stocking rate for livebearers is 50 mm (2 in.) in body length of fish to each 3.8 litre (1 gal) of aquarium capacity. If aeration and filteration are available, this capacity can be doubled. In general, poeciliid fishes do well at temperatures between 23–28 °C (73–82 °F), but thermostatic water heaters are recommended if the room temperature is not controlled.

Fishes of the genera *Xiphophorus* and *Poecilia* have broods of young about once a month throughout the year. The average brood size is between 20 and 40 fry, but may range from one to almost two hundred. Sexual maturity occurs at 3–14 mo of age depending upon the species, stock, genotype of the individual, food availability, water temperature,

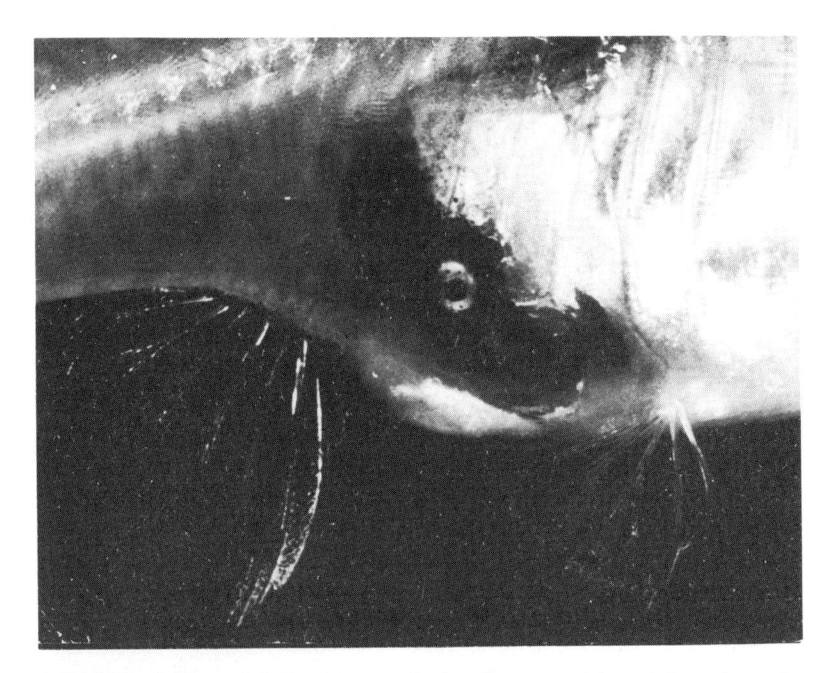

FIGURE 8 A close-up of the abdomen of a female guppy ready to deliver shows the eye of the unborn young.

and number of fishes per tank. Secondary sex characteristics are usually obvious.

2. Egglayers

Barbs from Asia and Africa, danios from Ceylon and India, and many South American characins spawn readily. Determination of sex is relatively easy; males may have longer and more pointed anal and dorsal fins and are more colorful and more slender than females. The males and females of these fishes should be held in separate aquaria until the females are gravid and the males show hyperactivity. A special spawning aquarium of about 8-litre (2-gal) capacity is prepared with clean, conditioned water and spawning grass made of Spanish moss or soft-leaved aquatic plants such as *Cabomba* and *Myriophyllum*. The ratio of sexes for breeding purposes is usually three males to two females. The females should be put into the aquarium the night before the males are introduced. The aquarium used for breeding should be constructed completely of glass or plastic so that the eggs can be seen when a light is directed through the bottom. The eggs of barbs, danios, and characins are tiny and clear. When spawning is complete the breeders should be removed since they devour their own spawn.

At a water temperature of about 27 °C (81 °F) the eggs normally hatch in 24 to 72 h. The young must be fed very small protozoans, followed by freshly hatched brine shrimp nauplii, *Artemia salina*. Instructions for hatching and feeding brine shrimp can be found in most general aquarium books (Axelrod and Schultz, 1955; Emmens, 1962; Axelrod, 1971; Axelrod and Vorderwinkler, 1972).

Many cichlids spawn easily (Goldstein, 1970). The male usually fertilizes the eggs, which are laid in a clutch of about twenty eggs to the string, as soon as the female leaves the spawning site. This sequence of events is periodically continued during the spawning season, resulting in the production of several hundred eggs. One advantage of this spawning behavior is that the exact fertilization time of any one egg can be determined.

The breeding pair of cichlids usually protect and aerate their eggs; however, the eggs should be removed and hatched separately from the parents. Cichlids breed all year, often at biweekly intervals, if the eggs are removed.

The average brood size for cichlids, 150 fingerlings, should be maintained at a stocking rate of about 4-litre (1-gal) capacity for every 25 mm (1 in.) of fish. As soon as the fingerlings are free-swimming, they usually can be fed newly hatched brine shrimp. The sexing and spawning habits of many cichlids have been well described (Axelrod and Shaw, 1967; Goldstein, 1970; Axelrod, 1971; Fryer and Iles, 1972; Goldstein, 1972).

3. Mouthbrooders

The mouthbrooders are mostly cichlids and a few anabantoids. Breeding fishes should be stocked at the rate of 25 mm (1 in.) of fish for each 4 litres (1 gal) of water. The eggs are deposited on the bottom of the aquarium, where they are fertilized. Either sex may brood the eggs during the several-weeks-long period of incubation. When the fingerlings hatch and become free-swimming, they venture out, but at the first sign of danger return to their parent's mouth (Baerends and Baerends-Van Roon, 1950; Axelrod and Shaw, 1967; Wickler, 1972).

4. Bubblenest Builders

Bubblenest builders are air breathers and, therefore, can be stocked at a rate of 0.1 m (4 in.) per 3.8 litres (1 gal) of water. Most gouramis and some catfishes build nests of mucus-covered bubbles on the surface of the water, while others build nests under the water (Axelrod and Shaw, 1967; Goldstein, 1972).

Of the bubblenest builders, the Siamese fighting fish has an involved mating process which has been described in great detail because of its unique courtship procedures and rearing of young (Gordon and Axelrod, 1968). The fingerlings have voracious appetites and require copious amounts of living infusoria before they reach a sufficient size to eat brine shrimp nauplii. Male Siamese fighting fish are pugnacious and must be kept separated, although the females may be kept together (Gordon and Axelrod, 1968; Goldstein, 1971).

IX Personnel

Care of experimental fishes does not differ fundamentally from that of other laboratory animals. Without basic training in the biology of fishes, it is difficult, if not altogether impossible, to take good care of experimental fishes. However, accuracy and trustworthiness are still more important than prior experience or training at the college level. In situations where many aquaria are in use, separate individuals should be responsible for tank and fish maintenance (feeding, cleaning, etc.).

The technical staff should have training and experience at the college level in biology, animal husbandry, and/or laboratory animal science with appropriate experience in the care of species involved. The staff should include one technician for general caretaking for every 25-50 aquaria of 40-55 litre (10-15-gal) size or larger, or every 25 ponds of 1/100-1/25 acre surface area, or every 50 ponds of lesser size depending upon the rate of rotation of use. Care and maintenance of experimental fishes should only be done by carefully trained technicians. These individuals may be recruited from the ranks of tropical fish hobbyists. Persons whose hobbies include aquarium fishes are more likely to be better qualified to take care of fishes in the laboratory. However, close supervision of the technical staff is required and fish health should be monitored regularly.

For the novice, information on the technical aspects of fish culture is available through state and federal fish and game offices or from the fish and wildlife resources of the land-grant and sea-grant colleges and universities. Frequently, extension specialists in fisheries and fisheries biology can be contacted for assistance. If these sources of information are not readily available, fish hobbyists and pet dealers may be consulted.

Many technical schools, colleges, and universities offer short training courses in laboratory animal science, and prospective employees should be encouraged to participate in these programs to better qualify for the job of caring for these experimental animals.

X Nutrition

Until 1930, the diets of fishes were formulated mostly on purely arbitrary or empirical considerations. In that year a special laboratory for research on nutrition of trout was established at Cortland, New York, where outstanding research was carried out on principles of fish nutrition (Cortland Hatchery Reports, 1933 to 1970). A similar laboratory for fish nutrition specializing in Pacific salmon nutrition was established in 1950 by the U.S. Fish and Wildlife Service at Cook, Washington.

Nutrition of pondfishes, catfishes in particular, is being studied at various laboratories. Much of this work is carried out at a federal laboratory located in Marion, Alabama (Dupree, 1966; Dupree and Sneed, 1966).

Nutrition of tropical aquarium fishes is based largely on experience and is described in textbooks on the maintenance and culture of aquarium fishes (Axelrod and Schultz, 1955; Emmens, 1962; Axelrod, 1971). Comprehensive literature on fish nutrition has been published (Halver, 1972, 1973).

A. DIETS FOR FRESHWATER EXPERIMENTAL FISHES

Since the nutrition of experimental freshwater fishes may influence research results, diets should be carefully selected by investigators. Unfortunately, most fishes used in research are fed one diet for a portion of their life cycles, then offered another ration prior to or during the experiment. When held in intensively managed facilities, fishes become conditioned, not only to their environment and maintenance schedules but also to the consistency and color of their food. Rainbow trout switched from a soft meat diet to a hard pelleted diet refused to feed until the hard pellets had been coated with meat, which was then gradually withdrawn from the diet (Phillips, 1970). Also, rainbow trout and bluegills maintained on a dark pelleted ration required a conditioning

TABLE 3 Percentages of Various Components in the Moist, Purified Diet Used To Feed Fish Pesticide Research Fishes[a]

Component	Percent of Various Components in Complete Diet
Casein	20.0
Gelatin	4.35
Dextrin	13.2
Mineral mix (Bernhardt–Tomarelli, modified)	2.00
Alphacel	3.85
Alpha-tocopherol (660 IU/kg)	0.1
Choline chloride 70%	0.5
Vitamin mix[b]	1.0
Fish oil	1.5
Corn oil	3.5
Total dry components	50.0
Water added	50.0[c]

[a] From Castell *et al.* (1972) the diet is fed to subadults and adults, whereas small fingerlings receive low-residue commercial feed.
[b] See Table 4 for quantities of various vitamins.
[c] In reality, water content of the dry mix is measured and an appropriate amount of water is added to give 50% water.
Source: Adapted from Brauhn and Schoettger (in press).

period before accepting light purified diet pellets (Phillips *et al.*, 1957). An abrupt ration change will induce undesirable stress in fishes and influence the initial results of experiments (Castell *et al.*, 1972). Fishes intended for toxicological research should be maintained on the ration to be used during research (Tables 3 and 4). A low-pesticide diet for routine fish maintenance and for pesticide-enriched rations used for research is acceptable (Brauhn, 1971). Experimental freshwater fishes fed this diet grow well, without apparent abnormalities (Brauhn and Schoettger, in press). Components of the diet given in Table 3 are commercially available in different states of purity. Therefore, in research requiring a high degree of reproducibility all components must be accompanied by information on their chemical analysis. More details on research diets are given by Halver (1972).

Many diets have been devised to maintain fishes for varying durations of time. Hunn *et al.* (1968) used a combination of eight diets, including live *Daphnia*, to feed bioassay fishes during a pretreatment holding period. Peterson *et al.* (1967) designed a meal-gelatin diet for aquarium fishes that contained meat to increase palatability. Formulations for typical diets used in fish culture in the United States, vitamin premixes, and a formula for a complete synthetic fish diet are furnished by Halver (1972). Salmonids were maintained on this diet without impairment of any functions for generations (Hashimoto and Okaichi, 1969).

Natural food organisms from lakes, streams, or ponds should not be considered for routine maintenance of experimental freshwater fishes because of possible contamination with industrial pollutants and pesticides, variation in supply from season to season, diversity of weight and size, and variation between species in protein, carbohydrate, and fat content. Natural foods may also be vectors of diseases that infect fishes. Diets fed at the hatchery can be the source of undesirable chemical residues in tissues of research fishes (Eisler, 1967; Pickering and Vigor, 1965). Although commercial dry diets provide adequate nutrition for rainbow trout, fathead minnows, channel catfish, and bluegills, special consideration may be required when feeding fishes destined for specific kinds of research. Residues of pesticides and industrial contaminants in several commercial fish diets were analyzed from 1970 through 1972 (Table 5). These residues were also analyzed in various common components of fish

TABLE 4 Components of Special Vitamin Mix Used in the Moist, Purified Diet

Vitamin	Vitamin Mix (g/kg)
Thiamine hydrochloride	3.20
Riboflavin	7.20
Niacinamide	25.60
Biotin	0.08
Calcium pantothenate (D)	14.40
Pyridoxine hydrochloride	2.40
Folic acid	0.96
Menadione	0.80
Vitamin B_{12} (1,000 μg/g)	21.336
i-Inositol (meso)	125.00
Ascorbic acid	60.00
Para-aminobenzoic acid	20.00
Vitamin D_2 (500,000 IU/g)	0.40
Vitamin A palmitate (500,000 IU/g)	2.50
Alphacel	716.124

Source: Adapted from Brauhn and Schoettger (in press).

TABLE 5 Pesticides and Other Contaminants in Five Commercial Diets Analyzed by the Fish Pesticide Research Laboratory from 1970 through 1972

		Residues (μg/g)					
						Phthalates	
Company	Number of Samples	DDT[a]	Dieldrin	Methoxychlor	PCB[b]	Diethylhexyl	n-Butyl
1	3	0.04	–[c]	–	1.4	–	–
		0.08	–	–	0.13	–	–
		0.11	–	–	0.3	–	–
2	1	–	–	–	0.7	2.0	–
3	4	< 0.005	–	–	0.065	–	0.22
		< 0.005	–	–	0.11	–	0.17
		0.20	–	1.20	0.28	–	–
		0.03	0.01	–	< 0.05	–	–
4	3	–	–	–	0.4	–	–
		–	–	–	1.6	–	–
		–	–	–	2.0	–	–
5	2	0.05	–	0.05	0.23	–	–
		0.11	0.01	–	< 0.05	–	–
6[d]		0.006	–	–	0.10	–	–

[a]DDT + DDD + DDE.
[b]A mixture of Aroclor 1232, 1242, 1254, and 1260.
[c]Not detected.
[d]FPRL research diet.
Source: Adapted from Brauhn and Schoettger (in press).

diets (Table 6). In the past, the Cortland diet, formulated from selected ingredients (Phillips *et al.*, 1957), was used.

B. FOOD

Many different commercial types of tropical fish food are available, and most of them are acceptable. However, the ingredients of commercial foods may be changed without notice, and this could affect various parameters that are studied. Commercial dried foods come in various forms, sizes, and formulations (e.g., powder, pellets, flakes). The preparation to be used depends upon the species and size of the animal and also upon the number of fishes (Gaudet, 1970; Halver, 1972). If a large number of fishes is maintained over several years, it may be advantageous to prepare one's own formula (Gordon, 1950).

Adult fishes should be fed at least once per day, always at the same time. How much food is added to each tank is learned by watching the fishes feed. All food (this applies to prepared or dried food) should be eaten within 10 min. Excess food is to be removed to prevent fouling. Many fishes will not feed on waterlogged food or partially decomposed food even when hungry.

Newborn and young aquarium fishes should be fed a minimum of three times daily with an appropriate diet. Good nutrition during the first few days of life is critical for an adequate growth rate. Fishes poorly treated or starved during the first few days after birth will be stunted for the rest of their lives. Automatic feeding devices are available (Ghittino, 1972).

Newborn fishes should never be placed in freshly cleaned tanks (unless the design of the experiment so stipulates). A conditioned, well-balanced aquarium contains many microscopic organisms on which fishes feed. These organisms are absent from newly cleaned tanks.

In addition to the standard diet it is advantageous to raise some live food as an added source of vitamins and other trace elements. Certain species of fishes will be indifferent to prepared food or avoid it altogether, while they will feed readily on live organisms.

C. *DAPHNIA*

Daphnia can be raised indoors either in regular aquaria or in large tubs. The larger the volume of water, the more dependable and productive the supply and harvest. *Daphnia* cultures can be placed into odd corners of a room where they do not use any valuable space. Lighting to promote

TABLE 6 Pesticides and Other Contaminants in Fat, Protein, Carbohydrate, Vitamin, and Mineral Sources of Fish Diets[a]

	Residues (μg/g)							
							Phthalates	
Diet Component	DDT[b]	Dieldrin	Chlordane	Lindane	BHC	PCB[c]	*n*-Butyl	Diethylhexyl
Fats								
Salmon oil								
maximum	1.31	–[d]	–	–	–	1.8	–	–
minimum	0.14	–	–	–	–	0.7	–	–
Cod oil								
maximum	1.89	1.10	0.22	–	0.19	13.1	–	–
minimum	–	–	–	–	–	0.7	–	–
Herring oil								
maximum	0.65	–	–	–	–	1.9	–	–
minimum	–	–	–	–	–	–	–	–
Menhaden oil								
maximum	1.09	–	–	–	–	4.7	–	< 0.5
minimum	–	–	–	–	–	–	–	–
Pollock oil								
maximum	–	–	–	–	–	> 32.0	–	–
minimum	Not determined							
Corn oil								
maximum	–	–	–	–	–	0.1	0.1	1.0
minimum	–	–	–	–	–	–	–	–
Linseed oil								
maximum	No detectable residues in all samples							
minimum								
Proteins								
Fish meal								
maximum	0.08	–	–	–	–	0.6	0.1	< 0.5
minimum	< 0.005	–	–	–	–	–	–	–
Extracted fish meal[e]								
maximum	–	–	–	0.01	–	2.9	–	–
minimum	–	–	–	–	–	–	–	–
Casein								
maximum	< 0.005	–	–	–	–	–	< 0.1	< 0.5
minimum	–	–	–	–	–	–	–	–
Gelatin								
maximum	0.028	0.06	–	–	–	0.1	–	7.0
minimum	–	–	–	–	–	–	–	–

Skim milk								
maximum	No detectable residues in all samples							
minimum								
Soybean meal								
maximum	–	0.03	–	–	–	0.1	–	0.6
minimum	–	–	–	–	–	–	–	–
Carbohydrates								
Dextrin								
maximum	No detectable residues in all samples							
minimum								
Wheat middlings								
maximum	0.05	< 0.01	0.03	0.14	0.02	–	0.2	–
minimum	–	–	–	–	–	–	–	–
Cornstarch								
maximum	< 0.005	–	–	–	–	0.1	–	–
minimum	–	–	–	–	–	–	–	–
Vitamins, Minerals, and Binders								
Bone meal								
maximum	0.03	0.01	–	–	–	–	0.4	< 0.5
minimum	–	–	–	–	–	–	–	–
Distillers' solubles								
maximum	0.005	0.01	–	–	–	0.3	–	–
minimum	–	–	–	–	–	0.1	–	–
Brewers' yeast								
maximum	–	–	–	0.03	–	–	–	–
minimum	–	–	–	–	–	–	–	–
Carboxymethyl cellulose								
maximum	–	–	0.72	–	–	–	–	–
minimum	–	–	–	–	–	–	–	–
Alphacel	No detectable residues in all samples							
Mineral mix	No detectable residues in all samples							
Minimum detection levels	0.005	0.01	0.1	0.001	0.001	0.1	0.1	0.5

[a]Data was gathered in spot checks by FPRL from 1970 through 1972.
[b]DDT + DDE + DDD.
[c]Mixture of Aroclor 1232, 1248, 1254, and 1260.
[d]Not detected in analyses.
[e]Includes East and West Coast marine fish meals for human and animal consumption.

Note: In addition to the above residues, the following compounds were present in diet components: aldrin in fish meal (maximum of 0.01 μg/g) and in distillers' solubles (maximum of 0.1 μg/g); endrin in salmon oil (maximum of 0.03 μg/g); methoxychlor in wheat middlings (maximum of 2.08 μg/g); heptachlor in bone meal (maximum of 0.04 μg/g); malathion in wheat middlings (maximum of 0.5 μg/g).

Source: Adapted from Brauhn and Schoettger (in press).

algal growth can be provided with a combination of fluorescent and ultraviolet lights such as GroLux.

1. Setting Up a Large *Daphnia* Culture

An uncovered fiberglass tank of about 400–500-litre (100–125-gal) capacity can be used to set up a large culture. It is filled with tap water and left standing for at least 10 days of conditioning. Two aerators are set up with the air stones about 50 mm (2 in.) above the bottom. After conditioning, two heads of lettuce that have been blanched in hot water are added to the tank. After 6 weeks the tank is ready to receive the *Daphnia.*

Daphnia can be procured either commercially (pet stores, biological supply houses), from another laboratory, or from ponds. Under no circumstances, regardless of how they are obtained, should the *Daphnia* be dumped into the tank. This is to prevent other microscopic undesirable organisms (e.g., copepods, planarias, leeches, hydra) from entering the tank. The *Daphnia* can be placed into another dish and under a 10X magnification can be picked up with an eyedropper and placed into conditioned water. These *Daphnia* are used to seed the tanks. Only a dozen *Daphnia* are needed to obtain a "bloom" of many thousands within a few days.

2. Maintenance of *Daphnia* Culture

The tank is fed daily one teaspoon of brewers' yeast (dead or active). Yeast can be obtained either from local bakeries or through health food stores. Once a month one head of blanched lettuce is added. When about 15 percent of the water has been lost through evaporation, it is replenished directly from the tap. The volume of conditioned water in the tub is sufficient to absorb the untreated tap water without detrimental effects. A water temperature of 20–22 °C (68–72 °F) is adequate for the *Daphnia* culture. A heavy layer of sediment will accumulate at the bottom; this can be stirred up once a week for a period of 1 min at the end of the day. Once a year three or four pails of this muck can be taken out, using the same net with which *Daphnia* are caught. The *Daphnia* culture is odorless and can easily be maintained for years. At higher temperatures there may be a bloom of protozoa or rotifers and undesirable odors may develop. Under these circumstances it is best to suspend all feeding for a day or two. The *Daphnia* population may occasionally become low in numbers, but it will not be lost. No fishes should be dropped into the tank, since they will reduce the size of the *Daphnia* population and perhaps eliminate it.

3. Feeding *Daphnia* to Fishes

Several hours after the aerators in the pool are turned off, the *Daphnia* will congregate near the surface and can be collected with a fine-meshed 0.15 X 0.23-m (6 X 9-in.) net rapidly drawn through the water. The *Daphnia,* which have high oxygen requirements, can be stored in well-conditioned aquarium water inside an observation dish; depletion of oxygen, however, will lead to their death in several hours. In unconditioned water the *Daphnia* will succumb within an hour or two.

4. Other Organisms

Apart from protozoa and rotifers mentioned above, small, red tubifex worms may be introduced into the pool. They form tight little bundles and accumulate along the wall of the pool. They can be collected with a net and fed in the same manner as *Daphnia*. These worms are excellent for newborn livebearers.

D. BRINE SHRIMP

Both the newborn brine shrimp nauplii and the adults are excellent fish food. The nauplii are especially valuable for raising juveniles. Nauplii are raised in the laboratory from brine-shrimp eggs that are available commercially from local tropical fish stores. As implied in the name, brine shrimp live in salt water. In freshwater tanks they will die within 2 to 3 h. Adult brine shrimp, good for larger fishes, can be obtained commercially in the frozen form or alive.

Brine-shrimp eggs are placed in a salt solution or in natural seawater when available. The water must be strongly agitated. Hatching time is temperature dependent. At 24 °C (75 °F) most eggs will hatch within 48 h. A 20-litre (5-gal) polyethylene bottle with a needle-type spigot near the bottom can be used to hatch up to 6 teaspoons of brine-shrimp eggs.

The air circulation is turned off and the container slightly inclined toward the front by pushing a 25-mm (1-in.) thick piece of wood underneath the back of the bottle. After 15 min the empty eggshells will float to the surface while the pink nauplii will accumulate at the bottom. The faucet is opened and the brine shrimp are collected in a fine-meshed net; the salt water is drained into a bucket. If salt water is readily available, the water drawn from the hatcher can be discarded each time. If it is reconstituted seawater, it may be reused for about a week. It is important to get complete separation between the nauplii and the empty eggshells; both are eaten by the fishes, but the eggshells are indigestible and may clog the fishes' intestines.

E. OTHER FOOD

1. Mosquito Larvae

Mosquitos will breed in virtually any stagnant water, including various man-made containers of sufficient size, during the warm months in the north and all year long in the south of the United States. No feeding is necessary since sufficient airborne debris supports microscopic organisms on which mosquito larvae feed. The larvae can be caught in the outdoor container with a net and fed similarly to adult brine shrimp or *Daphnia.* Alternatively, the "egg rafts" can be skimmed off the surface and placed in an aquarium. The eggs will hatch within a day or two and the larvae will reach appropriate size within a week.

2. Tubifex Worms

Tubifex worms are readily eaten by fishes. Those from aquarium stores can be obtained by the pint or quart and are of larger size than those grown in *Daphnia* pools. They come in tight bundles and can be kept in the laboratory for many days if placed in a shallow pan and let stand under cold, slowly running, or dripping tap water. Alternatively, the pan with water can be placed in the refrigerator. Strongly chlorinated water is toxic to tubifex worms.

F. DIETS FOR MARINE FISHES

Nutrition and feeding of saltwater fishes have not reached the level of sophistication achieved with freshwater fishes because culture of marine species is still in experimental stages. Nevertheless, many marine fish species are maintained for long periods with feeding based on trial and error (Halver, 1972; Bardach *et al.*, 1972).

Prepared diets, such as Chinook salmon mash and pureed spinach, are readily accepted by mullets, croakers, pinfish, sheepshead minnows, killifishes, and silversides. The drums and flounders are carnivores and must be fed live food, such as crustaceans, mollusks, and smaller fishes. Most marine fishes will eat shrimp bits, squid, live guppies, brine shrimp, *Daphnia,* fresh mussels removed from shells, and other foods of this type. Food should be provided at regular intervals (twice a day) and left in the aquarium 10–15 min; all uneaten food should then be removed. Overfeeding must be avoided because it causes rapid deterioration of the conditions in the tanks and may result in death of fishes.

XI Communicable Diseases and Their Control

It is not possible to describe in this text all the important fish diseases and to present in detail the methods of diagnosis, prophylaxis, and treatment. These subjects are described in other published texts (Wolf, 1966; Sindermann, 1966, 1970a,b; Hoffman, 1967; Conroy, 1968; Conroy and Herman, 1970; Kabata, 1970; Snieszko, 1970; Sarig, 1971; Bullock *et al.*, 1971; Bullock, 1971; Goldstein, 1971; Poppensiek, 1973; Hoffman and Meyer, 1974). Bibliographies are also available (Snieszko *et al.*, 1970; Wright, 1971).

Among the diseases caused by viruses the most important are channel catfish virus disease, eel papilloma also called cauliflower disease, fish pox, infectious hematopoietic necrosis (IHN), infectious pancreatic necrosis (IPN), lymphocystis disease, spring viremia of carp, and viral hemorrhagic septicemia (VHS). See Appendix B for further information.

So far, there is no effective treatment for viral diseases. Therefore, it is important to obtain experimental fishes from a reliable source. Any fishes imported or exported must be accompanied by a certificate of health (Code of Federal Regulations, 1970).

Fishes to be used in experiments should always be quarantined since there is no simple way to detect latent infections.

Fish eggs must be procured from disease-free stock to avoid indiscriminate dissemination of exotic diseases that may present a hazard to man and to fishes. See Scheel (1968) for a complete description of all species of *Nothobranchius, Aphyosemion,* and *Epiplatys.*

There are very few known fish pathogens and parasites that can be transmitted from fishes to humans. Some species of *Aeromonas, Pseudomonas,* and *Vibrio* can be pathogenic to humans. Localized and persistent infections with mycobacteria have been reported, and certain digenetic Trematoda and Cestoda can infect people. For these reasons it is recommended that usual safety precautions be observed during the handling of experimental fishes.

Fish immunology has been described in a recently published book (Anderson, 1974).

Appendix A: Detailed Description of Calculation of Loading Factors for Some Species of Fishes*

Haskell (1955) devised a method of correlating feeding rate and oxygen consumption to arrive at a maximum permissible weight of fish held per cubic foot of water. This method is described by Eq. (1).

$$\text{Wt of fish} = \frac{(\text{wt of feed/cubic ft of water}) \times 100}{\text{Percent body wt fed}}. \qquad (1)$$

FRESHWATER FISHES

The weight of feed per cubic foot of water is dependent upon the weight of fish per cubic foot of water and the percent body weight fed daily as determined from feeding tables developed by Deuel *et al.* (1952).

An example of this method is as follows: Experimental determination concluded that 3,520 rainbow trout with an average weight of 16.2 g was the maximum number that could be safely maintained at a density of 1.15 lb per cubic foot of 16.7 °C of water. There were 1.1 water exchanges per hour in the tank and the feeding rate was 6.1 percent of the total body weight daily. By substituting the above data in Eq. (1) the following relationship is obtained:

$$\text{Permissible wt} = \frac{6.98}{\text{Percent body wt}}.$$

The percent body weight fed will vary with water temperature and change in fish size, but the numerator of the equation is a constant for rainbow

*Source: taken from Lennon and Walker (1964).

trout at that location. Tank loading rates and the number of water exchanges are also assumed to remain constant.

Willoughby (1968) correlated expected fish metabolism and subsequent oxygen requirement with the established feeding rate to estimate the maximum carrying capacity of a rearing unit. The maximum permissible amount of food to be fed was first determined from the relationship given in Eq. (2).

$$(O^a - O^b)\frac{(5.45)}{100}(\text{gpm}) = \text{Maximum amount of food/day}, \qquad (2)$$

where O^a = Dissolved oxygen concentration (ppm) of incoming water;
O^b = Dissolved oxygen concentration (ppm) of outflowing water or the minimum safe oxygen concentration for the species being maintained;
5.45 = Metric tons of water in 1 gal/min flow for 24 h;
100 = Grams of oxygen required to metabolize 1 lb of a 1,200 cal/lb diet;
gpm = Incoming water flow rate in gal/min.

This method was used in the following example in which 10 gpm of 16 °C raw water with 9.2 ppm dissolved oxygen was supplied to rainbow trout. The rainbow trout minimum dissolved oxygen level is approximately 5.0 ppm (Doudoroff and Shumway, 1970). By substitution in Eq. (2), the following relationship is obtained:

$$\begin{aligned}\text{Maximum quantity of food/day} &= (9.2-5.0)\frac{(5.45)}{100}\frac{(10)}{1}\\ &= (4.2)(0.545)\\ &= 2.3 \text{ lb.}\end{aligned}$$

By substitution, the carrying capacity was determined for a holding tank containing rainbow trout fed at a rate of 6 percent of their body weight per day:

$$\begin{aligned}\text{Carrying capacity} &= \frac{\text{lb feed/day [from Eq. (2)]}}{\text{Percent body wt fed daily [from Deuel } et\ al. \text{ (1952)]}}\\ &= \frac{2.3 \text{ lb}}{0.06}\\ &= 38.3 \text{ lb.}\end{aligned}$$

Piper (1971) presented a method for determining the rainbow trout carrying capacity of rearing tanks that combined the methods of Haskell (1955) and Willoughby (1968). The feeding guide prepared by Buterbaugh and Willoughby (1967) was used to determine the relationship of fish length in inches and the percent body weight fed daily. The loading capacity was then determined by establishing the permissible weight of a given size of fish in pounds at a given water inflow rate. This relationship is expressed in Eq. (3):

$$F = \frac{W}{L \times I}, \tag{3}$$

where F = Loading factor
W = Known permissible weight of fish
L = Length of fish
I = Water inflow in gal/min.

Piper (1971) presented a table of loading factors dependent upon water temperature and altitude for trout and salmon, but the factor may be experimentally determined for Eq. (3). Once the loading factor is determined for a location, this remains constant. Also, the average length of fishes being maintained may easily be determined at periodic intervals. An example of this method for a holding unit having 90 lb of 4-in. rainbow trout safely maintained with 15 gal/min of water inflow would be as follows:

from (3) $F = \dfrac{90}{4\,(15)} = 1.5.$

This loading factor was then used to determine the maximum weight of 2-in. rainbow trout that was possible to maintain in a water flow of 5 gal/min.

From above: $1.5 = \dfrac{W}{2\,(5)}$
$W = 15$ lb fish.

Each presented method of determining the proper carrying capacity is adaptable to a variety of situations. However, these methods were developed for use with salmonid species only, and the results of these methods are approximations. Accuracy of estimating carrying capacity is increased by precise parameter measurements. The minimum dissolved oxygen concentration requirements for channel catfish, bluegill, and fathead minnows are lower than the 5.0 μg/g requirement of rainbow trout (Doudoroff and Shumway, 1970; Moss and Scott, 1961). Subsequently, we calculate carrying capacities of all holding tanks in terms of rainbow trout capacities. Preliminary testing of Piper's (1971) method with bluegill and largemouth bass (*Micropterus salmoides*) indicates that the relationship between the percent of body weight to feed daily and length of these fishes should be further investigated. Phillips (1970) presents further examples of the three methods discussed, a summary of the feeding tables of Deuel *et al.* (1952), and an explanation of other methods for estimating the carrying capacity of holding units.

MARINE FISHES

The carrying capacity of a small marine system can be calculated by a complex formula derived by Hirayama (1966). This formula is also found in the volume by Spotte (1970).

Appendix B: Fish Diseases and Their Control

TABLE B-1 Diseases of Fishes

Disease	Causative Agent	Control[a]
1. *Caused by Viruses*		
Channel catfish virus disease (disease of channel catfish in the United States)	Channel catfish virus	Avoidance of infection
Eel papilloma (cauliflower disease: stomatopapilloma)	Virus not yet identified	Same
Fish pox	Fish pox virus	Same
Infectious hematopoietic necrosis (disease of Pacific salmon species)	Infectious hematopoietic necrosis virus	Maintenance of water above 15 °C
Infectious pancreatic necrosis	Infectious pancreatic necrosis virus	Avoidance of infection. Fishes older than 6 mo are resistant
Lymphocystis disease (freshwater and marine fishes during summer months)	Lymphocystis virus	Avoidance of infection
Spring viremia of carp (carp in central Europe)	*Rhabdovirus carpio*	Avoidance of infection
Viral hemorrhagic septicemia (rainbow trout in Europe)	Egtved virus	Avoidance of infection
2. *Caused by Bacteria*		
Bacterial gill disease (mainly in hatchery salmonids)	Unidentified myxobacteria when fish kept in crowded conditions	1. Avoidance of crowding 2. Quarternary ammonia disinfectants 3. Diquat
Bacterial hemorrhagic septicemia (worldwide occurrence; also red leg disease of frogs)	*Aeromonas liquefaciens* (synonyms: *A. punctata, A. hydrophila*)	1. Avoidance of infection 2. Oxytetracycline or chloramphenicol orally

TABLE B-1 (continued)

Disease	Causative Agent	Control[a]
Coldwater disease (salmonids)	*Cytophaga psychrophila*	1. Oxytetracycline orally 2. Sulfamethazine or sulfisoxasole orally 3. Quarternary ammonia or diquat externally
Columnaris disease (worldwide occurrence; cottonmouth disease)	*Chondrococcus columnaris*	Same as for coldwater disease
Enteric bacteria (see also red mouth disease)	*Edwardsiella tarda*	1. Oxytetracycline orally 2. Pond sanitation
Fin rot	Multiple causes	Avoidance of crowding and water pollution
Fish furunculosis (worldwide except Australia and New Zealand; chiefly disease of salmonids)	*Aeromonas salmonicida*	1. Sulfamerazine 2. Oxytetracycline 3. Furoxone 4. Oral immunization (only under experimental conditions)
Gram-positive cocci	Rare	Avoidance of infection
Kidney disease (Dee disease) (in salmonids in North America and Scotland)	*Corynebacterium*; inadequately described (possibly *Listeria monocytogenes*)	1. Avoidance of exposure 2. Prophylaxis with sulfamethazine
Mycobacteriosis (universal; common in aquaria)	*Mycobacterium piscium, M. foruitum* and related	Avoidance of infection
Nocardiosis (same as mycobacteriosis, but much less common)	*Nocardia asteroides*	Avoidance of infection
Pasteurellosis (mostly marine fishes worldwide; not common)	*Pasteurella piscicida*	Avoidance of infection
Pseudomonas diseases (very similar to bacterial hemorrhagic septicemia)	*Pseudomonas fluorescens*; capsulated form very virulent	As in bacterial hemorrhagic septicemia
Red mouth disease (rainbow trout in the United States)	Red mouth enteric bacterium	1. Oral immunization 2. Oxytetracycline
Ulcer disease (chiefly brook trout in the United States)	*Hemophilus piscium*	Oxytetracycline and chloramphenicol orally

Vibriosis (worldwide; mostly in marine and estuarine environment; also in aquarium fishes)	*Vibrio anguillarium*	1. Oxytetracycline, nitrofurazone, and sulfamerazine orally 2. Furanace in bath externally 3. Oral immunization 4. Avoidance of infection

3. *Caused by Parasites*[b]

A. *Fungi*

Branchiomycosis	*Branchiomyces*	1. Avoidance of infection 2. Disinfection of ponds with quicklime or calcium cyanamide
Ichthyophonus disease	*Ichthyophonus hoferi* (*Ichthyosporidium hoferi*)	1. Avoidance of infection 2. Contaminated fish used as fish food
Saprolegniosis (as used in its broad meaning)	Species of *Saprolegnia, Achlya,* and others	1. Filtration of water 2. Disinfection of water 3. Malachite green 4. Formalin

B. *Algae–Dinoflagellates*

Oodiniosis (velvet disease)	*Oodinium*	1. Cleanliness of tanks 2. Copper sulfate 3. Malachite green

Diseases caused by green algae are not common.

C. *External Protozoans*

External protozoans are very common parasites of fishes causing frequent losses. The most-common are:

	Costia *Chilodonella* *Trichodina* *Amphiphrya* *Epistylis* *Trichophrya*	1. Avoidance of infection 2. Formalin 3. Copper sulfate 4. Malachite green 5. Methylene blue

TABLE B-1 (continued)

Disease	Causative Agent	Control[a]
D. *Intradermal Protozoans*		
Ichthyophthiriasis ("Ich")	*Ichthyophthirius*	1. Avoidance of infection 2. Killing of free-swimming forms with: a. Formalin b. Malachite green c. Copper sulfate 3. Temperature as high as fishes will tolerate for a few days
E. *Intestinal Protozoans*		
Inhabiting the lumen of the intestines		
Hexamitiasis	*Hexamita*	1. Enheptin 2. Cyzine 3. Carbarsone
F. *Systemic Protozoan Parasites*		
Systemic protozoan parasites are very common in nature. In cultured populations of fishes, control of infection by systemic protozoan parasites is feasible by observance of strict sanitary measures. Treatment of infected fishes is as yet not practical. The most common protozoan parasites are:		
	Cryptobia, Trypanosoma, Babesiosoma, Myxosporidae, Ceratomyxa in salmon	These parasites are seldom, if ever, transmitted directly from fish to fish. They can be transmitted by feeding infected fishes, by other parasites, as leeches or crustaceans, or require aging under natural conditions.
	Myxosoma, Myxobolus, Henneguya, Thelohanellus, Hofferelus, Nosema, Glugea, Plistophora, Thelohania	The only practical control is by avoidance. Chemotherapy is not sufficiently advanced but promising.

G. *External Trematode Worms or Monogenetic Flukes*

External trematode worms or monogenetic flukes are common and bothersome parasites. There are hundreds of species of which the most common belong to the following genera: *Gyrodactylus, Dactlylogyrus, Cleidodiscus, Benedenia.* They are easily transmitted directly from fish to fish by water. The best control is avoidance. They can be controlled with formalin, potassium permanganate, dylox, and bromex.

4. *Caused by Helminths*

The term helminth includes all parasites classified as Trematoda, Cestoda, Nematoda, and Acanthocephala. Larvae of those parasites are found in the flesh or viscera. Adult Cestoda, Acanthocephala, and Nematoda may inhabit the lumen of the intestinal tract. Helminths are not transmitted directly from fish to fish unless the intermediate host is another fish consumed by a predator fish. Birds, snails, and crustaceans serve as intermediate hosts.

Best control is by avoidance of the presence of intermediate hosts.

Some adult helminths inhabiting fish intestines can be treated successfully with oral drugs such as santonin, di-*n*-butyl tin oxide, or kamala.

Crustaceans are common, bothersome, and difficult to control. The most common are *Argulus, Ergasilus, Lernea, Achtheres,* and *Salmincola.*

There are freshwater and marine crustaceans. Some nonparasitics are intermediate hosts for helminths. Some copepoda bury deep in the flesh and cannot be affected by chemicals.

The most effective control chemicals are insecticides, such as bromex and dylox, added to water. Avoidance and selection of fishes free from crustacean parasites is the best control. Lesions caused by crustaceans may serve as portals of entry for bacteria and viruses. Crustaceans can also transmit agents of disease from fish to fish.

5. *Miscellaneous Diseases*

Fishes are subject to many neoplastic diseases and to nutritional diseases; are affected by sunlight, ultraviolet light, and fluorescent light; and often suffer from gas embolism if water is supersaturated with nitrogen or oxygen. Rapid changes in temperature are dangerous. Handling resulting in abrasions, loss of scales or mucus, or exposure to air may be either directly lethal or make the fishes more susceptible to infectious diseases.

---

[a] See also Table B-2.

[b] Viruses and bacteria are also parasites; however, it is customary to consider other parasites separately. Only the most common types of parasites (by genera only) are listed since numbers of parasites of fishes are large. For specific information readers are referred to the following detailed sources: Hoffman, 1967, 1970; Sindermann, 1970b; Conroy and Herman, 1970; Goldstein, 1971.

TABLE B-2 Chemicals Used Most Frequently for Control of Infectious Diseases of Fishes

Chemicals	Application
Acetic acid, glacial	Diluted in water: 1:500 for 30–60 s (dip) 1:2000 (500 ppm) as bath for 30 min
Acriflavine (trypaflavine)	5–10 ppm added to water from several hours to several days
Betadine (iodophore containing 1.0% of iodine in organic solvent)	100–200 ppm in water on the basis of iodine content by weight for fish egg disinfection
Bromex (dibrom, naled; a pesticide)	0.12 ppm added to water for indefinite period of time
Calcium cyanamide	Distributed on the bottom and banks of drained but wet ponds at a rate of 200 g/m^2
Calcium oxide (quicklime)	Distributed on the bottom and banks of drained but wet ponds at a rate of 200 g/m^2
Carbarsone oxide	Mixed with food at a rate of 0.2 percent; feeding for 3 days
Chloramphenicol (Chloromycetin)	1. Orally with food 50–75 mg/kg body wt/day for 5–10 days 2. Single intraperitoneal injection of soluble form 10–30 mg/kg 3. Added to water 10–50 ppm for indefinite time as needed
Chlortetracycline (Aureomycin)	10–20 ppm in water
Copper sulfate (blue stone)	For 1-min dip: 1:2000 (500 ppm) in hard water, 1 ml glacial acetic acid/litre
$CuSO_4$ anhydrous	0.25–2.0 ppm to ponds; quantity depends on hardness of water; hard water requires more
$CuSO_4 \cdot 5H_2O$ crystalline	
Cyzine (enheptin-A)	20 ppm in feed for 3 days
Diquat (patented herbicide, Ortho Co., contains 35.3% of active compound)	1–2 ppm of diquat cation, or 8.4 ppm as purchased added to water; treatment for 30–60 min; activity much reduced in turbid water
Dylox (dipterex, neguron, chlorophos, trichlorofon, foschlor)	0.25 ppm to water in aquaria and 0.25–1.0 ppm in ponds for indefinite period
Formalin (37% by weight of formaldehyde in water; usually contains also 12–15% methanol)	1:500 for 15-min dip 1:4000–1:6000 for 1 h 15–20 ppm to pond or aquarium water for indefinite period
Furazolidone (furoxone N.F. 180; N.F. 180 Hess & Clark) commercial products contain furazolidone mixed with inert materials	On the basis of pure drug activity: 25–30 mg/kg body wt/day up to 20 days orally with food

Furanace (P-7138) (made in Japan)	Added to water with fish to be treated at 1 ppm for several hours; toxicity to different fishes varies from 0.5 to 4.0 ppm (experimental drug)
Hyamine 1622 (Rohm & Haas Co., quarternary ammonium germicide available as crystals or in 50% solution)	1.0–2.0 ppm in water for 1 h
Hyamine 500 (as above)	(as above)
Iodophores	(see betadine and Wescodyne)
Kamala	Mixed with diet at a rate of 2%; feeding to starved fishes for 3 days
Malachite green	1:15,000 in water as a dip for 10–30 s
	1–5 ppm in water for 1 h (most often used as 5 ppm)
	0.1 ppm in ponds or aquaria for indefinite time
Methylene blue	1.0–3.0 ppm in water for 3–5 days
Neguvon	(see dylox)
Oxytetracycline (Terramycin)	50–75 mg/kg body wt/day for 10 days with food (Law requires that it must be discontinued for 21 days before fishes are killed for human consumption.)
Potassium permanganate ($KMnO_4$)	1:1000 (1000 ppm) for a 10–40 s dip
	10 ppm up to 30 min
	3–5 ppm added to aquarium or pond water for indefinite time
Quinine hydrochloride (or quinine sulfate)	10–15 ppm in water for indefinite time
Roccal (benxalkonium chloride, quarternary ammonia germicide; see also hyamine 500; sold as 10–50% solution)	1–2 ppm in water for 1 h; toxic in very soft water; less effective in hard water
Sodium chloride (table salt, iodized or not)	1–3% in water from 30 min to 2 h only for freshwater fishes
Sulfamerazine	200 mg/kg body wt/day with food for 14 days (Law requires that treatment must be stopped for 21 days before fishes are killed for human consumption.)
Sulfamethazine	100–200 mg/kg body wt/day depending on the type of food with which it is mixed; for prophylaxis reduce the quantity to 2 g/kg/day; length of treatment as recommended
Sulfisoxazole (Gantrisin)	200 mg/kg body wt/day with food
Terramycin	(see oxytetracycline)
Tin oxide, di-*n*-butyl	25 mg/kg body wt/day with food for 3 days
Wescodyne (iodophore containing 1.6% of iodine in organic solvent)	100–200 ppm in water on the basis of iodine content by weight for 15 min for fish egg disinfection

Glossary

ANADROMOUS Fishes that move from marine to fresh water.

BIOASSAY SIZE Fishes used in metabolic studies of drugs, metals, or chemical substances.

CATADROMOUS Fishes that move from fresh water to marine water.

CLUTCH The number of eggs laid in any one string or egg mass.

DIADROMOUS Fishes that move back and forth between fresh and marine waters (eel); includes anadromous and catadromous.

EURYHALINE Fishes that have a wide range in physiological tolerance to the osmotic changes caused by changes in salinity (eel; Atlantic salmon, *Salmo solar*; Sticklebacks, *Gasterosteus*; Killifishes, *Fundulus*).

FLOWAGE An overflow of a stream or impoundment onto the adjacent land; flooding.

GRAVID Full of eggs or embryos; having the body distended with mature eggs.

JUVENILE Any fishes from the time of birth or hatching until they are of sufficient age to be sexually mature.

LABORATORY FISHES Fishes that have been bred and raised specifically for research use and have received the prescribed care and handling necessary to produce uniform, high-quality fishes of known age, free of diseases, and in a known state of nutritional health.

OVIPAROUS Fishes that shed their eggs and in which development of the embryos takes place outside the maternal body.

OVOTESTIS A gonad that contains both testicular and ovarian tissues.

OVOVIVIPAROUS Fishes that retain their eggs and in which development of the embryos takes place internally, the embryos depending little or not at all upon nutrients from the parent.

RESEARCH FISHES (See *laboratory fishes*)–Fishes that are differentiated from either "cultured" or "wild caught" by the fact that they are of known age, generally definable, and reared under known conditions with observation and monitoring of their health status.

SALINITY A measure of the total salts in a given weight of seawater expressed as 35 g/litre or 35 ppt.

SPAWN The fertilized eggs produced by one breeding pair.

SPAWNING The mating process in which the male fertilizes the ovum.

STENOHALINE Relatively intolerant to salinity changes.

VIVIPAROUS Fishes that produce, instead of eggs, living young that depend largely upon the parent for feeding.

References

Agranoff, B. W., R. E. Davis, and R. E. Gossington. 1971. Esoteric fish. Science 171:236.

Anderson, D. P. 1974. Fish Immunology. Book 4 *in* S. F. Snieszko and H. R. Axelrod, eds. Diseases of fishes. T. F. H. Publications, Neptune City, N.J. 239 p.

Atz, J. W. 1971. Aquarium fishes. Viking Press, N.Y. 110 p.

Axelrod, H. R. 1971. Breeding aquarium fishes, Book 2. T. F. H. Publications, Neptune City, N.J. 352 p.

Axelrod, H. R. 1973. Koi of the world, Japanese colored carp. T. F. H. Publications, Neptune, N.J. 240 p.

Axelrod, H. R., and L. P. Schultz. 1955. Handbook of tropical aquarium fishes. McGraw-Hill, N.Y. 718 p.

Axelrod, H. R., and S. R. Shaw. 1967. Breeding aquarium fishes, Book 1. T. F. H. Publications, Neptune City, N.J. 480 p.

Axelrod, H. R., and W. Vorderwinkler, eds. 1972. Encyclopedia of tropical fishes, 21st ed. T. F. H. Publications, Neptune City, N.J. 800 p.

Baerends, G. P., and J. M. Baerends–Van Roon. 1950. An introduction to the study of the ethology of cichlid fishes. Brill, Leiden, Holland. 250 p.

Bailey, R. M., Chairman. 1970. A list of common and scientific names of fishes from the United States and Canada, 3rd ed. Am. Fish. Soc., Washington, D.C. Special Publ. No. 6. 150 p.

Bardach, J. E., J. H. Ryther, and W. O. McLarney. 1972. Aquaculture, the farming and husbandry of freshwater and marine organisms. Wiley-Interscience, N.Y. 868 p.

Bell, G. R. 1967. A guide to the properties, characteristics, and uses of some general anaesthetics for fish. Fish. Res. Bd. Can. Bull. 9 p.

Binkowski, F. P. 1972. Methods and techniques for collecting and maintaining alewives for biological research. Drum Croaker 13:21–27.

Brauhn, J. L. 1971. Fall spawning of channel catfish. Prog. Fish-Cult. 33:150–152.

Brauhn, J. L., and R. A. Schoettger. (in press). Acquisition and culture of research fish: rainbow trout, fathead minnows, channel catfish and bluegills. U.S. Environmental Protection Agency (Research Series), Washington, D.C. 45 p.

Brown, M. E., ed. 1957. The physiology of fishes: Vol. I. Metabolism. Academic Press, N.Y. 447 p.

Brown, M. E., ed. 1957. The physiology of fishes: Vol. II. Behavior. Academic Press, N.Y. 526 p.

Brungs, W. A. 1973. Continuous-flow bioassays with aquatic organism procedures and applications. Pages 117–126 *in* Biological methods for the assessment of water quality. American Society for Testing Materials: ASTM STP 528.

Bullock, G. L. 1971. Identification of fish pathogenic bacteria. Book 2B *in* H. R. Axelrod and S. Snieszko, eds. Diseases of fishes (series). T. F. H. Publications, Neptune City, N.J. 41 p.

Bullock, G. L., D. A. Conroy, and S. F. Snieszko. 1971. Bacterial diseases of fishes. Book 2A *in* H. R. Axelrod and S. F. Snieszko, eds. Diseases of fishes (series). T. F. H. Publications, Neptune City, N.J. 151 p.

Bureau of Sport Fisheries and Wildlife, Fish Pesticide Research Lab. 1972. Progress in sport fishery research. Toxicological Research, Bur. Sport Fish. and Wildl., Resour. Publ., U.S. Dep. Interior, Washington, D.C. 18 p.

Burrows, R. E., and B. D. Combs. 1968. Controlled environments for salmon propagation. Prog. Fish-Cult. 33:150–152.

Buterbaugh, G. C., and H. Willoughby. 1967. A feeding guide for brook, brown, and rainbow trout. Prog. Fish-Cult. 29(4):210–215.

Carlson, A. R. 1973. Induced spawning of largemouth bass (*Micropterus salmoides* [Laepedel]). Trans. Am. Fish. Soc. 102(2):442–444.

Carlson, A. R., and J. G. Hale. 1972. Successful spawning of largemouth bass, *Micropterus salmoides,* under laboratory conditions. Trans. Am. Fish. Soc. 101(3): 534–542.

Castell, J. D., R. O. Sinnhuber, J. H. Wales, and D. J. Lee. 1972. Essential fatty acids in the diet of rainbow trout (*Salmo gairdneri*): growth, feed conversion, and some gross deficiency symptoms. J. Nutr. 102(1):77–86.

Clark, J. R., and R. L. Clark, eds. 1971. Sea water systems for experimental aquariums. T. F. H. Publications, Neptune City, N.J. 192 p.

Code of Federal Regulations. Title 50—Wildlife and Fisheries. Chapter I, Part 13. Importation of Wildlife and Eggs Thereof. Revised January 1, 1970. Government Printing Office, Washington, D.C.

Conroy, D. A. 1968. Partial bibliography on the bacterial diseases of fish. An annotated bibliography for the years 1870–1966. Food and Agric. Organ. Fish. Tech. Paper No. 73, FRs/T73, CB-Bibliography, Rome, Italy. 75 p.

Conroy, D. A., and R. L. Herman. 1970. Textbook of fish diseases. (Transl. and revision of Taschenbuch der Fischkrankheiten by Erwin Amlacher.) T. F. H. Publications, Neptune City, N.J. 302 p.

Cortland Hatchery Reports, Nos. 1–39. 1933–1970. (annu. publ.). U.S. Fish and Wildl. Serv. and State of N.Y. Dep. Environ. Conserv., Cortland.

Culley, D. D., Jr., and Ferguson, D. E. 1969. Patterns of insecticide resistance in the mosquito-fish, *Gambusia affinis.* J. Fish. Res. Bd. Can. 26:2395–2401.

Cullum, L., and J. T. Justus. 1973. Housing for aquatic animals. Lab. Anim. Sci. 23:126–129.

Davis, H. S. 1953. Culture and diseases of game fishes. University of California Press, Berkeley and Los Angeles. 332 p.

Deuel, C. R., D. C. Haskell, D. R. Brockway, and O. R. Kingsbury. 1952. The New York State feeding chart. N.Y. State Conserv. Comm., Fish. Res. Bull. No. 3. 61 p.

Dollar, A. M., and M. Katz. 1964. Rainbow trout brood stocks and strains in American hatcheries as factors in the occurrence of hepatoma. Prog. Fish-Cult. 26(4): 167–174.

Doudoroff, P., and D. L. Shumway. 1970. Dissolved oxygen requirements of freshwater fishes. United Nations, Rome, Italy. FAO Fish. Tech. Paper No. 86. 291 p.

Dupree, H. K. 1966. Vitamins essential for the growth of channel catfish. Bur. Sport Fish. and Wildl., Washington, D.C. Tech. Paper No. 7. 12 p.

Dupree, H. K., and K. E. Sneed. 1966. Response of channel catfish fingerlings to different levels of major nutrients in purified diets. Bur. Sport Fish. and Wildl., Washington, D.C. Tech. Paper No. 9. 21 p.

Eisler, R. 1958. Some effects of artificial light on salmon eggs and larvae. Trans. Am. Fish. Soc. 87:151–162.

Eisler, R. 1967. Acute toxicity of zinc to killifish, *Fundulus heteroclitus.* Chesapeake Sci. 8(4):262–264.

Ellender, R. D., C. L. Armour, and B. J. Camp. 1971. Analysis of ammonia-nitrogen in seawater aquaria. J. Fish. Res. Bd. Can. 28:788–789.

Ellender, R. D., S. McConnell, G. W. Klontz, and L. C. Grumbles. 1973. Maintenance of gulf teleosts in closed artificial recirculated seawater systems (CARSWS) I. Problems and progress. Southwest. Vet. 26(3):201–206.

Emmens, C. W. 1962. Keeping and breeding aquarium fishes. T. F. H. Publications, Neptune City, N.J. 202 p.

Flickinger, S. A. 1971. Pond culture of bait fishes. Colorado State Univ. Bull. 478A. 39 p.

Fry, F. E. J. 1957. The aquatic respiration of fish. Pages 1–63 *in* M. E. Brown, ed. The physiology of fishes. Vol. 1. Academic Press, N.Y.

Fryer, G., and T. D. Iles. 1972. The cichlid fishes of the great lakes of Africa: their biology and evolution. Oliver & Boyd, Edinburgh. 641 p.

Gall, G. A. E. 1972. Rainbow trout brood stock selection program with computerized scoring. The Resources Agency of Calif., Inland Fisheries Administrative Report No. 72-9. 20 p.

Gaudet, J. E., ed. 1970. Report of the 1970 workshop on fish feed technology and nutrition. Government Printing Office, Washington, D.C. 207 p.

Ghittino, P. 1969. Piscicoltura e ittiopatologia: Vol. 1. Piscicoltura. Rivista de Zootech., Torino, Italy. 333 p.

Ghittino, P. 1972. The diet and general fish husbandry. Pages 539–650 *in* J. E. Halver, ed. Fish nutrition. Academic Press, N.Y.

Goldstein, R. 1970. Cichlids. T. F. H. Publications, Neptune City, N.J. 256 p.

Goldstein, R. 1971. Diseases of aquarium fishes. T. F. H. Publications, Neptune City, N.J. 126 p.

Goldstein, R. 1972. Anabantoids and other gouramis. T. F. H. Publications. Neptune City, N.J. 160 p.

Gordon, M. 1950. Fishes as laboratory animals. Pages 345–449 *in* E. J. Farris, ed. The care and breeding of laboratory animals. John Wiley & Sons, N.Y.

Gordon, M. 1953. The use of fishes in the evaluation of heredity in atypical pigment cell growth. Trans. N.Y. Acad. Sci., Ser. II 15:192–195.

Gordon, M., and H. R. Axelrod. 1968. Siamese fighting fish. T. F. H. Publications, Neptune City, N.J. 64 p.

Guidice, J. J. 1966. An inexpensive recirculating water system. Prog. Fish-Cult. 28(1):28.

Halver, J. E., ed. 1972. Fish nutrition. Academic Press, N.Y. 713 p.

Halver, J. E., Chairman. 1973. Nutrient requirements of trout, salmon, and catfish. National Academy of Sciences–National Research Council, Washington, D.C. 57 p.

Halver, J. E., and I. A. Mitchell, eds. 1967. Trout hepatoma research conference papers. U.S. Fish Wildl. Serv. Res. Refs. 70:1–199.

Hashimoto, Y., and Okaichi, T. 1969. Vitamins as nutrients for fish. Translation published by F. Hoffmann. La Roche & Co. Ltd., Basel, Switzerland. 40 p.

Haskell, D. C. 1955. Weight of fish per cubic foot of water in hatchery troughs and ponds. Prog. Fish-Cult. 17(3):117–118.

Hazard, T. D., and R. E. Eddy. 1951. Modification of the sexual cycle in brook trout (*Salvelinus fontinelis*) by control of light. Trans. Am. Fish. Soc. 80:158–162.

Heighway, A. J., ed. 1973. Fish farming international. Fishing News (Books) Ltd., West Byfleet, Surrey, England. 152 p.

Henderson, C., Q. H. Pickering, and J. M. Cohen. 1959. The toxicity of synthetic detergents and soaps to fish. Sewage Ind. Wastes 31(3):295–306.

Henderson, C., and C. M. Tarzwell. 1957. Bioassays for control of industrial effluents. Sewage Ind. Wastes 29:1002.

Hiatt, R. W. 1963. World directory of hydrobiological and fisheries institutions. American Institute of Biological Sciences, Washington, D.C. 320 p.

Hickling, C. F. 1962. Fish culture. Faber & Faber, London. 295 p.

Hinshaw, R. N. 1973. Pollution as a result of fish cultural activities. U.S. Environmental Protection Agency, Office of Research and Monitoring, Washington, D.C. Ecol. Res. Ser. No. 9. 209 p.

Hirayama, K. 1966. Studies on water control by filtration through sandbed in a marine aquarium with closed circulating system. IV. Rate of pollution of water by fish, and the possible number and weight of fish kept in an aquarium. Bull. Jap. Soc. Sci. Fish. 32:20–26.

Hoar, W. S., and D. J. Randall, eds. 1969–1971. Fish physiology. 6 Vols. Academic Press, N.Y. and London.

Hoffman, G. L. 1967. Parasites of North American freshwater fishes. University of California Press, Berkeley. 486 p.

Hoffman, G. L. 1970. Control and treatment of parasitic diseases of freshwater fishes. Fish Disease Leaflet No. 28, Bur. Sport Fish. and Wildl., Washington, D.C. 7 p. (A detailed monograph by Hoffman is being edited for publication.)

Hoffman, G. L., and F. P. Meyer. 1974. Parasites of freshwater fishes: a review of control and treatment. T. F. H. Publications, Neptune City, N.J. 224 p.

Hoffman, R. E. 1884. Artificial seawater for aquaria. U.S. Fish Comm. Bull. 4(30):465–467.

Holmes, W. N., and E. N. Donaldson. 1969. The body compartments and the distribution of electrolytes. Pages 1–89 *in* W. S. Hoar and D. J. Randall, eds. Fish physiology. Vol. I. Academic Press, N.Y. and London.

Huet, M. 1973. Textbook of fish culture. Fishing News (Books) Ltd., West Byfleet, Surrey, England. 436 p.

Hunn, J. B., R. A. Schoettger, and E. W. Whealdon. 1968. Observations on the handling and maintenance of experimental fish. Prog. Fish-Cult. 30:164–168.

Jacobs, K. 1971. Livebearing aquarium fishes. Studio Vista Limited, Blue Star House, Highgate Hill, London N 19. 460 p.

Kabata, Z. 1970. Crustacea as enemies of fishes. Book 1 *in* H. R. Axelrod and S. Snieszko, eds. Diseases of fishes (series). T. F. H. Publications, Neptune City, N.J. 171 p.

Kallman, K. D. 1970. Different genetic basis of identical pigment patterns in two populations of platyfish, *Xiphophorus maculatus.* Copeia (3):472–487.

Kallman, K. D., and J. W. Atz. 1966. Gene and chromosome homology in fishes of the genus *Xiphophorus.* Zoologica 51:107–135.

Klontz, G. W., and L. S. Smith. 1968. Methods of using fish as biological research subjects. Pages 323–386 *in* William I. Gay, ed. Methods of animal experimentation. Vol. III. Academic Press, N.Y.

Leitritz, E. 1960. Trout and salmon culture. Calif. Dep. of Fish and Game, Sacramento. Fish. Bull. 107. 169 p.

Lennon, R. C. 1967. Selected strains of fish as bioassay animals. Prog. Fish-Cult. 29(3):129–133.

Lennon, R. C., and C. R. Walker. 1964. Laboratories and methods for screening fish-control chemicals. Bur. Sport Fish. and Wildl., Circ. No. 185, Washington, D.C. Investigations in Fish Control No. 1. 15 p.

Lerner, I. M. 1954. Genetic homeostasis. Oliver & Boyd, Edinburgh. 134 p.

Lindsey, J. R., Chairman. 1969. Genetics in laboratory animal medicine. National Academy of Sciences, Washington, D.C. 66 p.

Lorenz, K. 1963. On aggression. Harcourt, Brace, and World, N.Y. 306 p.

Lyman, J., and R. Fleming. 1940. Composition of sea water. J. Mar. Res. 3:132–146.

Lyubitskaya, A. I. 1956. The effect of various parts of the visible area of the spectrum on the development stage of fish embryos and larvae. Zool. Zhur. 35(12): 1873–1886; Ref. Zhur., Biol., 1959, No. 91441; Biol. Abstr., 1963, 44(5), Abstr. No. 19785.

Lyubitskaya, A. I., and E. A. Dorofeeva. 1961. Effect of visible light, ultraviolet rays and temperatures on metamerism of fish bodies. Report No. 2. Effect of ultraviolet rays on survival and metamerism of the body in *Esex lucius L.* and *Acerina cernua L.* Zool. Zhur. 40(7):1046–1057; Ref. Zhur. Biol., 1962. No. 17114; Biol. Abstr., 1963, 42(3), Abstr. No. 8571.

Marking, L. 1966. Evaluation of *p,p'*-DDT as a reference toxicant in bioassays. U.S. Dep. Interior Resour. Publ. No. 14., Washington, D.C. Investigations in Fish Control No. 10. 10 p.

Martin, F. D. 1968. Intraspecific variation in osmotic abilities of *Cyprinodon variegatus* Lacepede. Ecology 49:1186–1188.

Mawdesley-Thomas, L. E. 1971. Neoplasia in fishes: a review. Cur. Topics Comp. Pathobiol. 1:88–170.

May, R. C. 1971. An annotated bibliography of attempts to rear the larvae of marine fishes in the laboratory. NOAA Technical Report NMFS SSRF-632, Government Printing Office, Washington, D.C. 24 p.

McFarland, W. N., and G. W. Klontz. 1969. Anesthesia in fishes. Fed. Proc. 28(4): 1535–1540.

McKim, J. M., G. M. Christensen, J. H. Tucker, D. A. Benoit, and M. J. Lewis. 1973. Effects of pollution on freshwater fish. J. Water Pollut. Control Fed. 45:1370–1407.

Morton, K. E. 1956. A new mechanically adjustable multi-size fish-grader. Prog. Fish-Cult. 18(2):62–67.

Moss, D. D., and D. C. Scott. 1961. Dissolved oxygen requirements for three species of fish. Trans. Am. Fish. Soc. 90(4):377–393.

Neuhaus, O. W., and J. E. Halver, eds. 1969. Fish in research. Academic Press, N.Y. 311 p.

Nigrelli, R. F. 1953. The fish in biological research. Trans. N.Y. Acad. Sci., Ser. II 15:183–186.

Parisot, T. J. 1967. A closed recirculated seawater system. Prog. Fish-Cult. 29(3):133–139.

Perlmutter, A. 1951. Possible effect of lethal visible light on year-class fluctuations of aquatic animals. Science 133(3458):1081–1082.

Perlmutter, A., and E. White, 1962. Lethal effect of fluorescent light on the eggs of the brook trout. Prog. Fish-Cult. 24(1):26–30.

Peterson, E. J., R. C. Robinson, and H. Willoughby. 1967. A meal–gelatin diet for aquarium fishes. Prog. Fish-Cult. 29(3):170–172.

Phillips, A. M., Jr. 1970. Trout feeds and feeding. Chapter 5, part 3, section B of Manual of fish culture. Government Printing Office, Washington, D.C. 49 p.

Phillips, A. M., Sr., H. A. Modoliak, D. R. Brockway, and R. R. Vaughn. 1957. Cortland Hatchery Report No. 26 for the year 1956. N.Y. Conserv. Dep., Albany. Fish. Res. Bull. No. 21. 93 p.

Pickering, Q. H., and W. M. Vigor. 1965. Acute toxicity of zinc to eggs and fry of the fathead minnow. Prog. Fish-Cult. 27(3):153–158.

Pickford, G. E., and J. W. Atz. 1957. The physiology of the pituitary gland of fishes. N.Y. Zool. Soc., N.Y. 613 p.

Piper, R. G. 1971. Know the proper carrying capacities of your farm. Am. Fishes and U.S. Trout News 15(1):4–6.

Poppensiek, G. C., Chairman. 1973. Aquatic animal health. National Academy of Sciences–National Research Council, Washington, D.C. 46 p.

Pruginin, Y., and E. W. Shell. 1962. Separation of the sexes of *Tilapia nilotica* with a mechanical grader. Prog. Fish-Cult. 24(1):37–40.

Pyle, E. A. 1966. The effect of grading on the total weight gained by three species of trout. Prog. Fish-Cult. 28(1):29–33.

Regier, H. A., and W. H. Swallow. 1968. An aquarium temperature control system for field stations. Prog. Fish-Cult. 30(1):43–46.

Richardson, I. D. 1972. Rearing of marine fish: problems of husbandry, nutrition, and disease. Symp. Zool. Soc. London 30:341–357.

Sarig, S. 1971. The prevention and treatment of diseases of warmwater fishes under subtropical conditions, with special emphasis on intensive fish farming. Book 3 *in* H. R. Axelrod and S. Snieszko, eds. Diseases of fishes. T. F. H. Publications, Neptune City, N.J. 127 p.

Scheel, J. 1968. Rivulins of the old world. T. F. H. Publications, Neptune City, N.J. 480 p.

Schnarevich, I. D. 1960. On the effect of different parts of the visible spectrum upon the growth and development of the brook trout (*Salmo trutta fario*) during the embryonic period. Nauch Ezhegodnik Za 1959 G. Chernovitskii Univ. Biol. Fak. 428–430. From: Ref. Zhur. Biol., 1964, No. 4138 (translations); Biol. Abstr., 1965, 46(1), Abstr. No. 2407.

Scott, K. R. 1972a. Temperature control system for recirculation fish-holding facilities. J. Fish. Res. Bd. Canada 29:1082–1083.

Scott, K. R. 1972b. Comparison of the efficiency of various aeration devices for oxygenation of water in aquaria. J. Fish. Res. Bd. Canada 29(11):1641–1643.

Segedi, R., and W. E. Kelley. 1964. A new formula for artificial sea water. Pages 17–19 *in* J. R. Clark and R. L. Clark, eds. Sea water systems for experimental aquariums. T. F. H. Publications, Neptune City, N.J.

Shelbourne, J. E. 1971. The artificial propagation of marine fish. T. F. H. Publications, Neptune City, N.J. 83 p.

Shell, E. W. 1966. Comparative evaluation of plastic and concrete pools and earthern ponds in fish-cultural research. Prog. Fish-Cult. 28(4):201–206.

Sindermann, C. J. 1966. Diseases of marine fishes. Ad. Mar. Biol. 4:1–89.

Sindermann, C. J. 1970a. Principal diseases of marine fish and shellfish. Academic Press, N.Y. 369 p.

Sindermann, C. J. 1970b. Bibliography of diseases and parasites of marine fish and shellfish. Tropical Atlantic Biologic Laboratory Informal Report No. 11. 44 p.

Snieszko, S. F. 1970. Immunization of fishes: a review. J. Wildl. Dis. 6:24–30.

Snieszko, S. F., F. T. Wright, G. L. Hoffman, and K. Wolf. 1970. Selected fish disease publications in English. U.S. Dep. Interior, Bur. Sport Fish. and Wildl. Fish Disease Leaflet No. 26. 7 p.

Snow, J. R. 1962. A comparison of rearing methods for channel catfish fingerlings. Prog. Fish-Cult. 24:112–118.

Spotte, S. H. 1970. Fish and invertebrate culture—water management in closed systems. Wiley-Interscience, N.Y. 145 p.

Spotte, S. H. 1973. Marine aquarium keeping. John Wiley & Sons, N.Y. 171 p.

Stephan, C. E., and D. I. Mount. 1973. Use of toxicity tests with fish in water pollution control. Pages 167–177 *in* Biological methods for the assessment of water quality. American Society for Testing Materials: ASTM STP 528.

Steucke, E. W., Jr., L. H. Allison, R. G. Piper, R. Robertson, and J. T. Bowen. 1968. Effects of light and diet on the incidence of cataract in hatchery-reared lake trout. Prog. Fish-Cult. 30(4):220–226.

Strickland, J. D. H., and T. R. Parsons. 1968. A practical handbook of seawater analysis. Fish. Res. Bd. Can. Bull. 167. 311 p.

Tamura, T. 1966. Marine aquaculture (Transl. from Japanese). National Technical Information Service, U.S. Dep. Commerce, Springfield, Va. PB–1041T. 1255 p.

Umbreit, W. W., and E. J. Ordal. 1972. The use of goldfish as an experimental animal in the undergraduate microbiology laboratory. Am. Soc. Microbiol. News 38:92–98.

Wickham, D. A., J. B. Eagle, and F. Hightower, Jr. 1971. Apparatus for controlling ambient light cycles in experimental environments. Trans. Am. Fish. Soc. 100(1):128–136.

Wickler, W. 1972. Breeding behavior of aquarium fishes. T. F. H. Publications, Neptune City, N.J. 192 p.

Willoughby, H. A. 1968. A method for calculating carrying capacities of hatchery troughs and ponds. Prog. Fish-Cult. 30(3):173–175.

Wolf, K. 1966. The fish viruses. Adv. Virus Res. 12:35–101.

Wright, F. T. 1971. List of reference sources for students of fish diseases. U.S. Dep. Interior, Bur. Sport Fish. and Wildl. Fish Disease Leaflet No. 33. 11 p.

To assist ILAR in determining the usefulness of this guide, please complete and return this questionnaire to the Institute of Laboratory Animal Resources, National Academy of Sciences, 2101 Constitution Avenue, Washington, D.C. 20418.

QUESTIONNAIRE

Name ____________________

Institution ____________________

Department ____________________

1. Specific area of scientific interests?

2. What portions of the guide did you find particularly useful?

   a. Which sections did you find difficult to use?

   b. Were any sections contrary to your experience or misleading?

   c. Were any sections too vague or overly explicit?

3. Suggestions for the improvement of future issues:

   a. Format:

   b. Contents:

   c. Other:

4. Purpose for which guide was obtained:

   General reference ______
   Specific requirement for information ______
   Other (specify) ______